2nd International Conference on Aeronautical, Aerospace and Mechanical Engineering (AAME 2019)

IOP Conference Series: Materials Science and Engineering
Volume 638

Prague, Czech Republic
26 – 28 July 2019

ISBN: 978-1-5108-9840-0
ISSN: 1757-8981

Printed from e-media with permission by:

Curran Associates, Inc.
57 Morehouse Lane
Red Hook, NY 12571

Some format issues inherent in the e-media version may also appear in this print version.

Institute of Physics
Dirac House, Temple Back
Bristol BS1 6BE UK

Phone: 44 1 17 929 7481
Fax: 44 1 17 920 0979

techtracking@iop.org

Additional copies of this publication are available from:

Curran Associates, Inc.
57 Morehouse Lane
Red Hook, NY 12571 USA
Phone: 845-758-0400
Fax: 845-758-2633
Email: curran@proceedings.com
Web: www.proceedings.com

Table of contents

Volume 638

2nd International Conference on Aeronautical, Aerospace and Mechanical Engineering (AAME 2019)

26–28 July 2019, Prague, Czech Republic

Accepted papers received: 4 September 2019
Published online: 11 October 2019

Preface

Preface

Delegate List

Conference Committees

Peer review statement

Papers

Aerospace Engineering

Extending Blade-Element Model to Contra-Rotating Configuration

Ohad Gur.....1

Hydraulic control system of mechanical energy stored type powder high velocity compaction machine

Dongdong You, Qian Zhu, Yanghui Zhu and Fenglei Li.....9

A study on decompresses heat pump dryer for drying of shiitake mushrooms at medium temperature

Y J Goh, Y H Kim, B Lee, H C Jung, S R Lee and K M Kim.....14

Ball on the plate balancing control system

Michael M Kopichev, Anton V Putov and Anastasia N Pashenko.....20

Research on the friction pair of the space high-power and long-life rolling ring

Chengshan Liu, Xiuqun Chen and Zhiyuan Qian.....26

A novel anti-lock braking strategy for high speed light weight dual-motor electric drive tracked vehicles

Zhaomeng Chen, Xiaojun Zhou and Yuanhe Liu.....35

A Position Synchronization Control for Dual Redundant Electro-mechanical Actuation in Flight Control System

D Y Wang, D Q Li, J Chen and S C Zheng.....41

Sliding Mode Variable Structure Control Strategy Applied on Position Servo System of EMA

D Y Wang, S J Gao, M Yan and S C Zheng.....53

Research on automatic assembly technology for final assembly of helicopter fuselage

Q D Xiao, L A Zhang, L Gan and T Cheng.....61

Study on abnormal behavior of human in safety accidents of special equipment

Hongqi Luo and Xia Liu.....70

Material Science

Investigation on Polymer Composite Materials Wear Reinforced by Microparticles of Jatropha Curcas L. Waste

Miroslav Müller, Petr Valašek, Martin Tichý, Alessandro Ruggiero and Roberto D'Amato.....77

Numerical Investigation of Hybrid of Eglass and Basalt Fiber Reinforced Epoxy Tube Pressurized Internally

Thamir A. D. M. S. Almula, Ahmad Hassan Khuder, Mohd Yazid Yahya and Amran Ayob.....85

Evaluation of the Slow Release Oxygen Properties and pH Control Ability of Oxygen Slow-releasing Materials

Zehao Chen, Pengfei Guo, Hailong Tian, Ying Wang and Chenghong Feng.....94

Agri-residual Filler Based Bio-Composites: *Synergistic Effect of Organo-sulfide Coupling Agent and Alkali Treatment on Mechanical Properties*

Firoz Alam Faroque, Saanil Kakati, Sanchita Bandhopadhyay-Ghosh and Subrata Bandhu Ghosh.....100

Optimization of Process Parameters for Closed-die Forging Based on Finite Element Simulation

Wei Guo and Jin Zou.....106

Analysis of optimal profile modification amount of spur gear under disturbance of backlash and center distance

Enen Shen, Yumei Hu, Zao He, Huan Zhang and Yanghan Min.....111

Study on the compacting device of combination for ridge-till and no-till planters

Hongqi Luo, Xia Liu, Jiatong Qin and Guangwei Zhao.....121

Effect of Ru on the concentration distribution of elements and creep properties of nickel-based superalloy

Shuang Liang and Zhixin Liu.....127

Micro-electrochemical evaluation of α- or β-type titanium alloys

Junhui Lin, Jiafei Lu, Zhenhua Dan, Yi Ding and Hui Chang.....134

IOP Conf. Series: Materials Science and Engineering **638** (2019) 011001 doi:10.1088/1757-899X/638/1/011001

Preface

Aeronautical, Aerospace and Mechanical Engineering is a key ingredient of current advanced technologies supporting modern society. The tremendous rapid growth of these fields makes them highly interdisciplinary and then inevitably demands their mutual interactions. Under these circumstances, 2019 the 2nd International Conference on Aeronautical, Aerospace and Mechanical Engineering (AAME 2019) was organized as a series of successful international conferences held in Malaysia(2018).

The main goal of AAME 2019, which will be held from 26 to 28 July 2019, in CULS Prague, Czech, is to address latest original results in aeronautical, aerospace and mechanical engineering, including both theoretical advances and practical implementations, which are becoming more and more popular in industry and in our daily lives. AAME 2019 will provide a premier interdisciplinary platform for scientists, researchers, industry leaders, engineers and educators throughout the world to present and discuss the most recent innovations, trends, concerns, as well as practical challenges encountered, and streamline solutions in the fields of aeronautical, aerospace and mechanical engineering.

Towards this end, the Technical Programme Committee has assembled an excellent programme comprising of 4 excellent Keynote Speeches from renowned scientists from the world, 2 parallel technical sessions comprising of 19 presentations. A total of 19 full papers were selected from 35 original contributions after a rigorous review process.

On behalf of the Organising Committee, I wish to thank the keynote speakers and authors of selected papers for their outstanding contributions. I would also like to thank members of the organizing committee, anonymous reviewers for their great efforts. Without their contribution, dedication and commitment, we would not have achieved so much. We sincerely hope that you will find the AAME 2019 beneficial and fruitful for your professional development. We also hope that you will enjoy our hospitality and will have an enjoyable and memorable time in Prague.

Conference Organizing Chair, AAME 2019

Assoc. Prof. Petr Valášek,

Czech University of Life Science Prague, Czech Republic

IOP Conf. Series: Materials Science and Engineering **638** (2019) 011002 doi:10.1088/1757-899X/638/1/011002

Paper ID	First name	Surname	Institution
ME003	Ohad	Gur	Israel Aerospace Industries, Israel
ME006	Dongdong	You	South China University of Technology,
ME012	Kyoung Min	Kim	Chonnam National University, South
ME013	Michael M	Kopichev	Saint Petersburg Electrotechnical
ME019	Chengshan	Liu	Aerospace System Engineering
ME020	Zhaomeng	Chen	Zhejiang University, China
ME026	Deyi	Wang	Department National Key Laboratory of
ME027	Deyi	Wang	Department National Key Laboratory of
ME0003	Petr	Valášek	CULS Prague, Czech republic
ME0004	Thamir	Almula	Northern Technical University, Iraq
ME0005	Ying	Wang	Beijing Normal University, China
ME0006	Firoz Alam	Faroque	Manipal University Jaipur, India
ME029	Qingdong	Xiao	AVIC Manufacturing Technology
ME031	Hongqi	Luo	Beijing Technology and Business
ME033	GUO	Wei	Jiangxi Academy of Sciences, China
ME036	Enen	SHEN	Chongqing University, China
ME040	Hongqi	Luo	Beijing Technology and Business
ME041	Shuang	Liang	Yingkou Institute of Technology, China
ME044	Junhui	Lin	Nanjing Tech University, China

IOP Conf. Series: Materials Science and Engineering **638** (2019) 011002 doi:10.1088/1757-899X/638/1/011002

E-mail address
ogur@iai.co.il
youdd@scut.edu.cn
kkm@chonnam.ac.kr
mkopichev@gmail.co
liucsbuaa1125@163.c
chernzm@zju.edu.cn
deyiwang@vip.qq.co
deyiwang@vip.qq.co
valasekp@tf.czu.cz
thamir_own@yahoo.c
yingwang@bnu.edu.c
firozalam.faroque@ja
fenghuo4000@126.co
lhqfrq@126.com
guowei053@163.com
enenshen@163.com
lhqfrq@126.com
70080952@qq.com
zhenhuadan@njtech.e

IOP Conf. Series: Materials Science and Engineering **638** (2019) 011003 doi:10.1088/1757-899X/638/1/011003

Conference Committees

Conference General Chair

Prof. Vincent Lee Chieng Chen, Curtin University, Sarawak, Malaysia

Conference Organizing Chair

Assoc. Prof. Petr Valášek, Czech University of Life Science Prague, Czech Republic

Program Chairs

Prof. Ing. Miroslav Müller, Czech University of Life Science Prague, Czech Republic

Prof. Hamid Reza Karimi, Polytechnic University of Milan, Italy

Technical Committee Chair

Prof. Roberto D´Amato, Universidad Politécnica de Madrid, Madrid

Technical Committees

Asst. Prof. Nolan Concha, FEU-Institute of Technology, Philippines

Prof. Francesco Zirilli, Sapienza Università di Roma, Italy

Asst. Prof. Kamran Shavezipur, Southern Illinois University Edwardsville, USA

Asst. Prof. Wei (Will) Wang, University of Cincinnati, USA

Assoc. Prof. Tönük Ergin, Middle East Technical University, Turkey

Assoc. Prof. Eriki Ananda Kumar, TVETI, Ethio-China University, Ethiopia

Assoc. Prof. K. Kadirgama, Universiti Malaysia Pahang, Malaysia

IOP Conf. Series: Materials Science and Engineering **638** (2019) 011003 doi:10.1088/1757-899X/638/1/011003

Assoc. Prof.Rusdi Nur, Politeknik Negeri Ujung Pandang, Indonesia

Assoc. Prof. Dr. Yamuna Munusamy, Universiti Tunku Abdul Rahman, Malaysia

Prof. Dr. Hariati Taib, Universiti Tun Hussein Onn Malaysia, Malaysia

Assoc. Prof. Tzong-Hann Shieh, Feng Chia University, Taiwan

Dr. Ramasamy, Universiti Malaysia Pahang, Malaysia

Dr. Muhamad Mat Noor, University Malaysia Pahang, Malaysia

Dr. Simon Winberg, University of Cape Town, South Africa

Dr. Hafiz Muhammad Ali, University of Engineering and Technology, Pakistan

Prof. Ranganath M Singari, Delhi Technological University, India

Prof. Mirza Jahanzaib, University of Engineering and Technology, Pakistan

IOP Conf. Series: Materials Science and Engineering **638** (2019) 011004 doi:10.1088/1757-899X/638/1/011004

Peer review statement

All papers published in this volume of *IOP Conference Series: Materials Science and Engineering* have been peer reviewed through processes administered by the proceedings Editors. Reviews were conducted by expert referees to the professional and scientific standards expected of a proceedings journal published by IOP Publishing.

IOP Conf. Series: Materials Science and Engineering **638** (2019) 012001 doi:10.1088/1757-899X/638/1/012001

Extending Blade-Element Model to Contra-Rotating Configuration

Ohad Gur[1]
[1] IAI – Israel Aerospace Industries, Ben-Gurion Airport, Lod 70100, Israel

Abstract. Blade element model is one of the most popular models for rotating-wing analysis and design. Many legacy tools are based on blade-element and capable of analysing only a single-rotating configuration. The current scope suggests a simple and effective method for extending such applications to contra-rotating configurations. The model based on the classic axial-momentum theory and assumes rigid wake. The method is validated versus results from propeller, hovering rotor, and prop-fan experiments. The experiments present both single and dual-rotating wing configuration, which share the same blade geometry and test facilities. This enables the isolation of the single- versus dual-rotating configuration influences, thus validating the suggested model with high level-of-confidence. The validation shows that the same trends of the single-rotating model appears in the dual-rotating analyses. Moreover, the difference between single and dual rotating configuration is replicated accurately compared to the test results. This proves the validity of the suggested model extension. The model validation also exhibits the differences between the accuracy of the forward and rear rotating disk in the contra-rotating configuration. The rear disk contributes by the no-swirled wake and exhibits accurate results compared to the forward disk which operate similar to a single rotating wing, hence suffers from the same inaccuracies.

1. Introduction

Blade-element model (BEM) is used extensively for rotating-wing analysis and design. This includes propellers, open-rotors, wind turbines, axial compressors, fans, etc. The advantage of BEM over many others, elaborate and accurate models, such as CFD based analyses, is the combination of simple model which contains accurate 2-D aerodynamic database for the blade's cross-sections. In that way, BEM is very simple and consumes very low computer resources. The complex effects of viscosity, compressibility, and stall, are introduced through the 2-D aerodynamic database. This is also the main draw-back of BEM – the requirement of valid, high level-of-confidence database, which suites large range of operating conditions, i.e. various Reynolds and Mach number. If such database exists, for axial design condition BEM exhibits excellent match to experimental results. [1]
These advantages enable researchers to use BEM for prop-fan and open-rotor analysis and design. In most cases, the validation of the model is limited, although the model is then used for vast range of cases. The current effort tries to cover these gaps.

2. Contra-rotating blade-element model

The single-rotating, SR, configurations are very common in various apparatus such as propeller, wind turbine, fan, etc. A single-rotating propeller with an axial free stream velocity, V, is replaced with an infinitesimal thin actuator-disk which introduces axial and circumferential momentum to the flow. The additional momentum is resulted with axial and circumferential induced velocity components, W_a and W_t, respectively. The source of these induced velocities is the propeller thrust and torque. According to Newton's third law, the thrust and torque that acts on the propeller is reacted by the same thrust and torque which acts on the flow. Thus, the air is accelerated to the axial direction due to the thrust and the wake is swirled due to the torque, i.e. axial and circumferential induced velocity is resulted. Note that while the flow passes the actuator disk, the axial velocity is continuous while the circumferential velocity is incontinuous and experiences a "jump" from zero to W_t.
A typical blade-element model is depicted as the rotating propeller's blade is sliced to small blade-elements, each dr width. The velocity components on the blade sections are summed resulting with total velocity, U, and local angle of attack, α. The blade-element acts as a 2-D wing section, thus it produces lift and drag forces, dL and dD, respectively. Summing all blade-element lift and drag components is resulted with the total thrust and torque which acts on the propeller.
In the current effort, the above blade-element model is extended to a contra-rotating case. Left part of Figure 1 shows an equivalent actuator-disks model. Each of the two actuator-disks acts the same manner as the single-rotating case. The main difference is the induced velocities. Except of the axial and circumferential induced velocities which the disk induces on itself, the disks induce velocities on each other. The rear disk induces axial cross-induced velocity on the forward disk, $W_{a,1on2}$, and the forward disk induces on the rear disk both axial cross-induced velocity, $W_{a,1on2}$, and circumferential cross-induced velocity, $W_{t,1on2}$. Note that the rear disk does not induce circumferential induced velocity on the forward disk – only the wake swirls

All of the above velocity components are taken into account in the contra-rotating blade-element scheme. It shows the forward and rear blade-elements which are described, each with its velocity diagram. The forward blade-element axial velocity components are the free stream, V, the self-induced axial velocity, $W_{a,1}$, and the axial cross-induced velocity which is induced by the rear disk, $W_{a,1,2on1}$. Both axial induced velocities act to the same direction. The forward circumferential components are the tangential speed due to the blade rotating, $\Omega_1 \cdot r$, and the circumferential self-induced velocity, $\frac{1}{2}W_{t,1}$. Note that the ½ factor on the self-induced circumferential velocity is due to the discontinuity of the swirl of the flow. Thus, the average ½ is taken for the blade-element velocity scheme.

The rear blade-element axial velocity components include the free stream velocity, V, the axial self-induced velocity, $W_{a,2}$, and the axial cross-induced velocity which the forward disk induces on the rear, $W_{a,1on2}$. The circumferential velocity components include the tangential velocity, $\Omega_2 \cdot r$, the self-induced circumferential induced velocity, $\frac{1}{2}W_{t,2}$, and the circumferential induced velocity which is induced from the forward on the rear disk, $W_{t,1on2}$. The circumferential induced velocities are opposite direction. While the self-induced $\frac{1}{2}W_{t,2}$ acts opposite of the propeller rotation, the cross-induced $W_{t,1on2}$ acts in the same direction of the rear-propeller rotation. In this way, the wake of the rear propeller experience very low rotational velocity which is one of the main advantages of contra-rotating configuration. This also implies on the zero torque of the contra-rotating configuration.

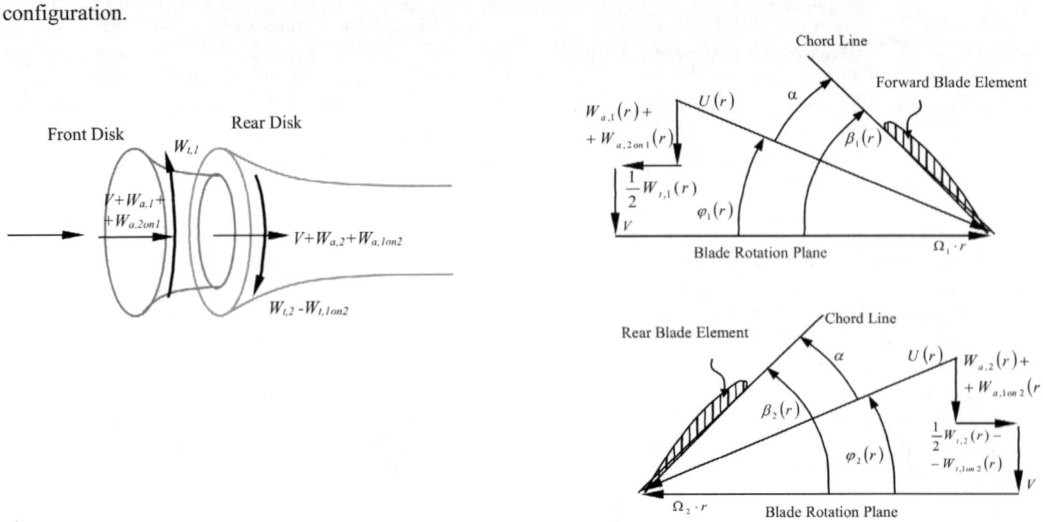

Figure 1. Contra- rotating propellers - schematic of the actuator disk and blade-element models depicting the velocity components.

Each of the disks in the contra-rotating problem is solved separately. Each solution influences the cross-induced velocities, $W_{a,2on1}$, $W_{a,1on2}$, $W_{t,1on2}$. A convergence is reached after iterative process. The solution of the contra-rotating problem is based on the solution of an isolated single-rotating disk. First the forward disk is solved and the cross induced velocities $W_{a,1on2}$ and $W_{t,1on2}$ are estimated according to the relative position of the rear disk. Then the rear disk is solved and the cross-induced velocity $W_{a,2on1}$ is calculated. The convergence of the disks solution is checked and then the convergence loop is closed. The convergence of the scheme is quick and generally after a few iterations both disks are converged.

The BEM for a single disk solution is the standard BEM as described in Ref. [2]. The main contribution of the current effort is the calculation of the cross-induced velocity components. This allows an existing single-rotating models to extend easily to a contra-rotating models. In what follows, the simplified model which uses to estimate the cross-induced velocities is presented.

3. Axial cross-induced velocities

The axial cross-induced velocities, $W_{a,1on2}$ and $W_{a,2on1}$ are calculated using the method described in Ref. [3]. It assumes that the axial induced velocity is constant in the radial coordinate. Although this assumption seems crude, it is quite accurate. Especially for optimized propellers, which are design to create radial uniform axial induced velocity. Using this assumption and simple momentum relation, the axial induced velocity, W_a, along the axial coordinate, x, can be found - Eq. (1).

$$\frac{W_a(x)}{W_a} = 1 + \frac{x}{\sqrt{x^2 + R^2}} \tag{1}$$

where W_a is the induced velocity on the disk plane, and $W_a(x)$ is the induced velocity in the axial coordinate x. R is the disk radius. Equation (1) suits negative x values, *i.e.* inflow, and positive values – wake. For high values of

x, Eq. (1) result with the axial induced velocity in the far wake. It is twice the value of the disk's (x=0) induce velocity. This match the classic result of the axial momentum theory presented by Glauert.[4]
Due to mass flux continuity, the inflow and wake radius, $R(x)$ can be found using Eq. (2).

$$\frac{R(x)}{R}=\sqrt{\frac{V+W_a}{V+W_a(x)}} \tag{2}$$

$R(x)$ is the radius of the inflow or wake according to the sign of the axial coordinate, x.
The assumption of constant induced velocity along the radial coordinate enables using the simple momentum model which relates the axial induced velocity, W_a, and the produced thrust, T. [4]

$$W_a=-\frac{V}{2}+\sqrt{\left(\frac{V}{2}\right)^2+\frac{T}{2\rho A}} \tag{3}$$

where A is the disk area and ρ is the air density.
Equation (3) can be written using the standard parameters of advance ratio, J, and thrust coefficient, C_T as depicted in Eq. (4).

$$\frac{W_a}{nD}=-\frac{J}{2}+\sqrt{\left(\frac{J}{2}\right)^2+\frac{2C_T}{\pi}} \tag{4}$$

where D is the disk diameter and n is the rotational speed in revolutions-per-second, rps, units. Advance ratio, J, and the thrust coefficient, C_T, power coefficient, C_P, and propeller efficiency, η, are defined in Eq. (5).

$$J=\frac{V}{nD} \quad C_T=\frac{T}{\rho n^2 D^4} \quad C_P=\frac{P}{\rho n^3 D^5} \quad \eta=\frac{J\cdot C_T}{C_P} \tag{5}$$

where P is the required power of the propeller.

4. Cross-induced circumferential induced velocity

The cross-induced circumferential induced velocity, $W_{t,1on2}$, model is based on simple momentum theory. Main assumption is that the wake is swirled at a constant rotation speed, Ω_t . Again, similar to the constant radial axial induced velocity, the rigid wake geometry fits optimized propeller which most well designed propellers are fit to. Thus, the circumferential induced velocity can be expressed using the wake rotational speed in Eq. (6).

$$W_t=\Omega_t\cdot r \tag{6}$$

To find the wake rotational velocity, the actuator disk is divided to co-centric annuluses, each introduces a circumferential momentum, thus increase the circumferential by W_t. this enable to calculate the contribution of each annulus to the propeller torque, dQ, according to Eq.

$$dQ=d\dot{m}\cdot W_t\cdot r \tag{7}$$

where $d\dot{m}$ is the mass flux through the annulus element. It is calculated using the axial velocity components through the disk plane – Eq. (8).

$$d\dot{m}=\rho\cdot(V+W_a)\cdot 2\cdot\pi\cdot r\cdot dr \tag{8}$$

The total torque, Q, is calculated using the integration of Eq. (7) and substituting Eqs. (6) and (8).

$$Q=\int_0^R \underbrace{\rho\cdot(V+W_a)\cdot 2\cdot\pi\cdot r\cdot dr}_{d\dot{m}}\cdot\underbrace{\Omega_t\cdot r}_{W_t}\cdot r \tag{9}$$

Assuming the axial induced velocity, W_a, is constant along the disk radii, the torque can be calculated and hence the wake rotational speed can be substantiated – Eq. (10).

$$\Omega_t=\frac{2\cdot Q}{\rho\cdot(V+W_a)\cdot\pi\cdot R^4} \tag{10}$$

Using the power coefficient, C_P, definition in Eq. (5), and Eq. (4), the wake rotational speed can be presented in a non-dimensional manner – Eq. (11).

$$\frac{\Omega_t}{\Omega}=\frac{8}{\pi^3}\frac{C_P}{J/2+\sqrt{(J/2)^2+2C_T/\pi}} \tag{11}$$

5. Biermann and Hartman test

Propeller performance maps for the Biermann and Hartman[5] test were calculated and are presented in Figure 2. It shows the performance maps for the 4-bladed propeller performance maps. Similar results, which are not presented here, were gained also for the 6 bladed propeller.
The maps are presented for 3 parameters; thrust coefficient, C_T, power coefficient, C_P, and propeller efficiency, η, as function of the advance ratio, J - all are defined in Eqs. (5).
The Biermann and Hartman's propellers shared the same blade geometry and had diameter of 120" (3.05 m). The blade uses Clark-Y airfoil family. The airfoil 2-D aerodynamic database was substantiated using MSES.[6]

Then PropSim (same tool used to generate the results in Ref. [1]) is used to calculate the propeller map. Figure 2 contains data for pitch angles vary from 20 degrees to 60 degrees for the 4 bladed propeller.

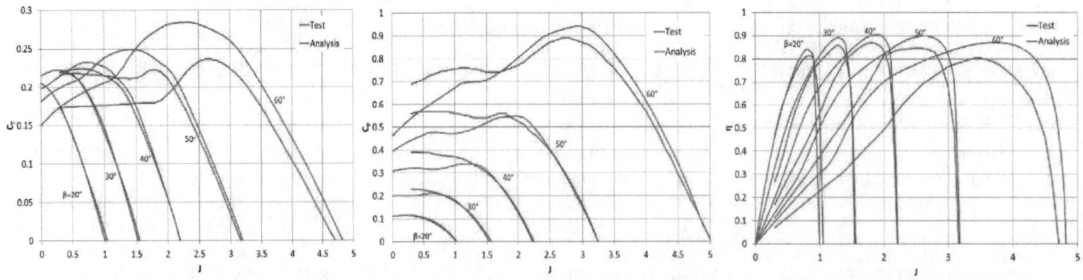

Figure 2. Test and analysis comparison for Ref. [5], 4 blades, single rotating propeller performance

Up to pitch angle of about 50 deg, the curves comparison exhibits a good agreement, both for the numerical values and for the curve shape and trends. The comparison is considered mainly to the linear regime, i.e. to the higher values of the advance ratio for each pitch angle. For example, for pitch angle of 40 deg. the linear regime is for advance ratio of J=1.5 and above. For this regime, efficiency comparison shows a maximum difference of about 5% efficiency, which is considered good. The thrust and power coefficient slopes shows a very good agreement. For low advance ratio, comparison of blade-element method is due to fail, mainly due to the high angle of attack which the blade sections experience. This was discussed thoroughly in Ref. [7] and a correction model is incorporated to the sectional 2-D aerodynamic database, used for the blade-element model. Still, some irregularities appear for the very low advance ratio which is caused due to the stalled airfoil database, as created by MSES.

In most cases MSES over-predicts the maximum lift coefficient [8],[9] thus the thrust coefficient also being over-predicted. The power coefficient comparison, at the lower advance ratio regime is fair, i.e. drag is being well estimated by MSES.

For high pitch angles, above 50 degrees, the comparison is deteriorated. Again, this statement is relevant for the high advance ratio *i.e.* linear regime. The efficiency difference reaches up to 10%, although the slopes and general trends of the thrust and power coefficients remain accurate.

Figure 3 shows the results for the contra-rotating propeller configurations. Note that the pitch setting was not the same for the forward and rear propellers and the pitch as stated by Biermann and Hartman is presented on the performance charts.

Figure 3 shows the performance of the 4-bladed propeller, thus the front and rear disk has 2 blades each, and total of 4 blades, i.e. same disk solidity as the single rotating case. Similar results exist also for the 6 bladed propeller with 3 blades for the front propeller and 3 for the rear.

The results are given again for the three performance parameters; thrust coefficient, C_T, power coefficient, C_P, and propeller efficiency, η. The test results included separate measurement of the power coefficient for the forward and the rear propellers, thus an additional comparison is given for these measurements.

In general, the comparison of analyses to test results exhibits the same trends as for the single propeller comparison, presented in Figure 2. The thrust and power coefficients agrees well with the test results for pitch angles up to β=50 degrees. Above this pitch angle, the agreement deteriorates although the curve slopes and the pitch sensitivity exhibit good agreement. For low advance ratios, the agreement is poor due to the same reasons of the single-rotating case, as shown in Figure 2. The efficiency curves exhibit up to 5% difference for low pitch angles and this difference increase to about 10% for the high pitch angles. Again, the trends are accurate, thus the sensitivity to advance ratio and pitch angle show good agreement to the test results.

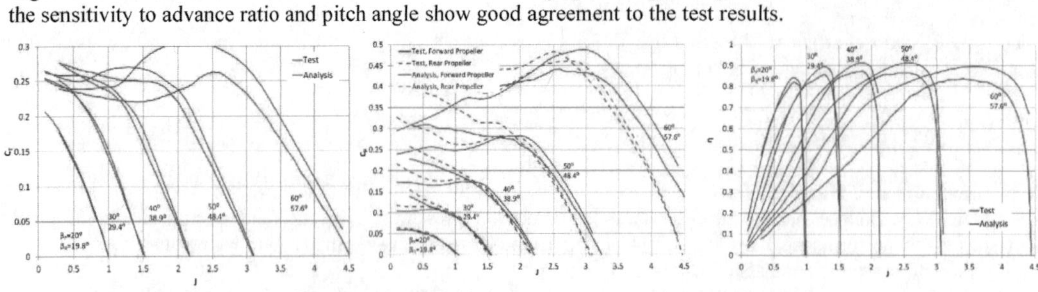

Figure 3. Test and analysis comparison for Ref. [5], 4 blades, contra-rotating propeller performance

IOP Conf. Series: Materials Science and Engineering **638** (2019) 012001 doi:10.1088/1757-899X/638/1/012001

Comparing the separate propeller measurements of the power coefficient teaches that the rear propeller agreement is excellent while the forward is less accurate. This result is correct also for high pitch angles well beyond the threshold of β=40 degrees which was concluded from the thrust and power coefficient curves. Therefore, the separate power coefficient results might hold some insight to this phenomenon.

The explanation lays in Figure 1 which depicts the blade-element of the forward and rear propeller along their velocities components. According to this description, the front propeller uses only a single estimation for the cross-induced velocity, $W_{a,2on1}$, while the rear, more accurate propeller uses two separate estimations for the axial and circumferential cross-induced velocities, $W_{a,1on2}$ and $W_{t,1on2}$, respectively. It is expected that the rear propeller will exhibit worse comparison. It is influenced by both axial and circumferential cross-induced velocities; both are crudely estimated. On the contrary, it seems that these cross-induced velocities rather improve the model accuracy.

The circumferential induced velocity for the rear propeller contains two contradicting components. The self-induced circumferential induced velocity, W_{t2}, acts in the opposite direction from the cross-induced component, $W_{t,1on2}$. In a way, the wake of a contra-rotating propeller might consider as a non-swirled wake, which implicate on the zero-torque that a contra-rotating propeller produces. Thus, the rear propeller acts in a non-swirled environment, which the only circumferential velocity component is the tangential velocity due to its own rotational speed. This causes the blade-element scheme of the rear propeller to be much accurate than the forward propeller, which includes the self-induced circumferential velocity, W_{t1}.

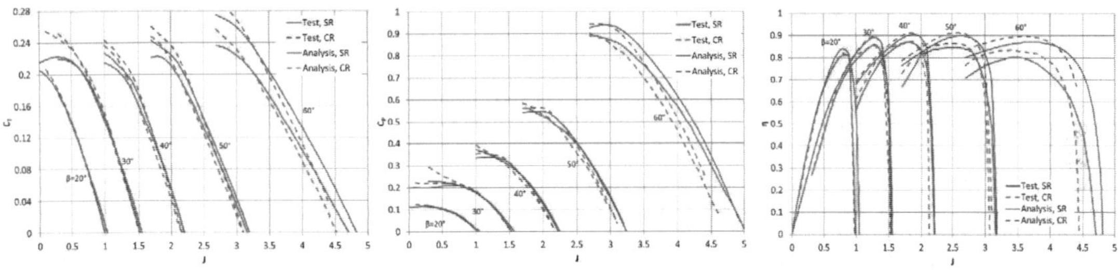

Figure 4. Test and analysis comparison for Ref. [5], 4 blades, SR & CR comparison

The accuracy of the circumferential induced velocity has major influence on the high pitch angles performance. Thus the forward propeller model is less accurate for high pitch angles. This explains the results of the separate power coefficient measurement and the exceptional agreement of the BEM model for the rear propeller. Moreover, it lays the blame for the disagreement of the total thrust and power coefficient, on the forward propeller, which acts similar to a single-rotating propeller and exhibits the same inaccuracies.

To further inspect the correctness of the proposed BEM for the contra-rotating case, a comparison for 5 pitch angles, β=20, 30, 40, 50, and 60 degrees, is presented in Figure 4. Here the curves are the same as in Figure 2 and Figure 3 for the 4-blade propeller. The purpose of Figure 4 is to present all curves; test, analysis, single rotating, SR, and contra-rotating, CR, on the same chart. Thus, the trends can be examined using the differences between the various curves. As stated above, very similar results were gained for the 6 bladed propeller.

The main trend is the difference between the SR and the CR. SR curves are solid while the CR curves are dashed. One can examine the gap between the solid and dashed curves for both test results (in red) and analyses (in blue). Both thrust and power coefficient exhibit decrease in value transferring from SR to CR. On the same time, the efficiency increases. As the pitch angle increases these trends become more prominent. The above observations repeat in the same manner for both the test and analysis results, thus proves the correctness of the current BEM model for the contra-rotating case.

6. Harrington test

Further validation of the contra-rotating model is given for hovering rotor, using the results by Harrington.[10] Harrington's test involves 2 different, hovering, 2-bladed, 25 ft (7.6 m) diameter rotors in a single-rotating configuration (SR) and contra-rotating, *i.e.* total of 4 blades. The rotor blades use NACA-4-digits symmetrical airfoils. The 2-D aerodynamic database, which was used to analyze these rotors, was substantiated from Ref. [11].

Figure 5 and shows the comparison of the rotor thrust and power coefficients, $C_{T,rotor}$, $C_{P,rotor}$, respectively, as defined in Eq. (12)

$$C_{T,rotor} = \frac{T}{\rho A V_{tip}^{2}} \qquad C_{P,rotor} = \frac{P}{\rho A V_{tip}^{3}} \tag{12}$$

where A is the disk area and V_{tip} is the blade's tip velocity.

These coefficients are equivalent to the propeller thrust and power coefficient, C_T and C_P, respectively as defined in Eqs. (5). Thus, the various coefficients can be related according to Eq. (13).

$$C_{T,rotor} = \frac{4}{\pi^3} C_T \quad C_{P,rotor} = \frac{4}{\pi^4} C_P \tag{13}$$

Figure 5 and Figure 6 show the rotors Figure-of-Merit, FM, which is the measure of the rotor quality at hover and is defined in Eq. (14).

$$FM = \frac{C_{T,rotor}^{1.5}}{C_{P,rotor}\sqrt{2}} \tag{14}$$

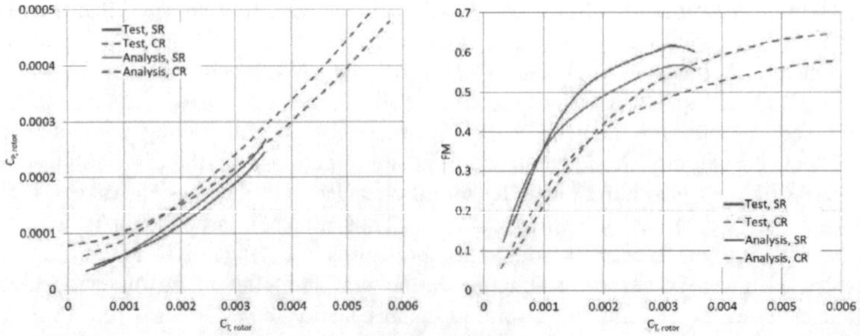

Figure 5. Test and analysis comparison for Ref. [10], Rotor I, SR & CR comparison

The results show good comparison between the experiment and analyses, both for the power to thrust coefficient ($C_{P,rotor}$ vs. $C_{T,rotor}$) chart and for the Figure-of-Merit to thrust coefficient (FM vs. $C_{T,rotor}$) chart. The trends are similar and values are close in within 5% error. The charts include both the SR and CR configuration. The difference between the two configurations is very similar comparing the test results and analysis. The power coefficient is increased by a constant value comparing the SR and CR configuration, while the curve slope and trend remains the same. As expected, the FM curves exhibit decrease from SR to CR configuration due to the increase of the required power for the same thrust coefficient – as depicted in Eq. (14). Similar to Biermann and Hartman test in Figure 4, the contra rotating model predicts well the differences between the SR and CR configuration, thus it is again validated.

7. Dunham et-al test

A third source for contra-rotating test results is available in Dunham et-al experiment.[12] They used the SR-2 propeller blades in a single-rotating 8 bladed propeller and 4+4 blades, contra-rotating configuration in a low-speed wind-tunnel, although the SR-2 was designed for high Mach numbers.[13] As a transonic propeller it has exceptionally thin blades, down to 2% thickness ratio, with relatively high chord to radius ratios. Sand et-al [14] contains a detailed model for the NACA-16 airfoil series, but it is not applicable for such thicknesses. Thus, a 2-D CFD analyses were conducted to substantiate the 2-D aerodynamic database, *i.e.* using EZair RANS code,[15] which was validated extensively.[16] As stated above, for the current case, the cross sectional Mach number is low due to the test conditions of this experiment.

The Dunham et-al experiment measured the performance of 3 pitch angles for the single-rotating, SR, configuration (β=30.45, 40.30, 50.15 degrees) and a single pitch angle (β=41.30 degrees) for the contra-rotating configuration. Unfortunately, for the CR configuration, only the thrust coefficient, C_T, was measured, thus somewhat lean comparison is available. Nevertheless, Figure 6 presents the thrust coefficient, C_T, power coefficient, C_P, and efficiency, η, of the SR-2 propeller.

The SR results exhibit good agreement for low pitch angle β=30.45 deg. As the pitch angle increases the agreement deteriorates - for the high pitch angle of β=50.15 degrees, the agreement is poor. For the intermediate pitch angle, β=40.3, the agreement is fair, with efficiency difference of up to 5%. The CR result seems to be good, although it is relevant only to the thrust coefficient. Still, the level of accuracy of the SR and CR results is similar, which again proves the correctness of the suggested contra-rotating analysis.

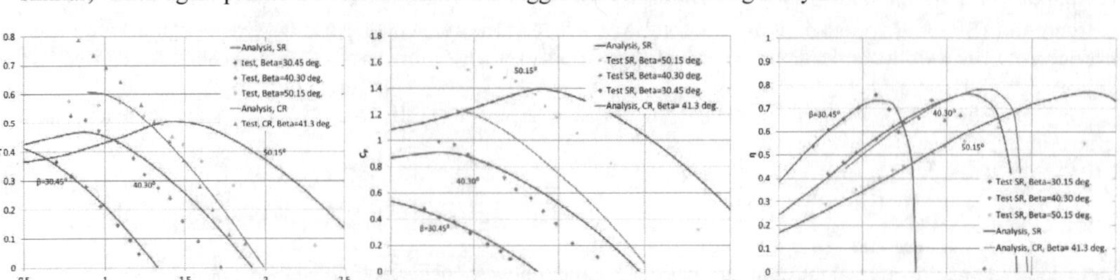

Figure 6. Test and analysis comparison for Ref. [12], SR & CR comparison

8. Summary and Conclusions

The paper exhibits the extension of a single-rotating blade-element model to case of contra-rotating configuration. While the single-rotating blade-element model is very common, the contra-rotating model is somewhat more complicated. The importance of the current paper is enabling the use of legacy single-rotating models and by simple means, extending it to contra-rotating applications. Moreover, the new contra-rotating model share the same accuracy level with the former, single rotating model.

The extension is based on iterative scheme of the single-rotating model with additional induced velocity components, which are induced by one rotating disk on the other. These cross-induced velocities are estimated using simple axial momentum theory for the axial cross-induced velocities and simple rigid rotated wake for the cross-induced circumferential velocity, which the forward rotating-wing induced on the rear.

Using this simple model, a well validated blade-element tool was used for the analyses of contra-rotating wing configurations. Three different test results were used as validation cases. These experiments contain propeller, hovering rotor, and prop-fan configurations. All tests used a single-rotating and dual-rotating configuration of the same blades, thus a focused view on the difference between the single and contra-rotating configuration can be conducted.

Each validation case was comprised of both single- and contra-rotating cases. The contra-rotating validations exhibit the same agreements and inaccuracies which were resulted for the single-rotated cases. Moreover, the differences between the single and dual-rotating configuration was replicated accurately by the analyses, and match in a good manner to the same difference exhibited by the experiment results. This proved the applicability of the contra-rotating extension model which used the same blade-element model for the single-rotating cases.

In general, for linear regime, non-stalled blades, the agreement is very good, up to 5% efficiency or figure-of-merit (for hovering cases). The deterioration of the agreement happens for the low advance ratio regime or high pitch cases. These inaccuracies are common both to the single and dual-rotating configuration, thus it is an inherit inaccuracy of the blade-element model.

In one of the validation cases, the forward and rear rotating-wings were measured separately. This enables to learn that the forward disk analysis suffers from inaccuracies more than the rear disk. The latter shows very good agreement between analyses and test data. This is due to the low circumferential induced velocity which the rear disk experiences. The circumferential induced model is one of the reasons for inaccuracies of single-rotating blade-element model, for high pitch cases. Thus, the rear rotating-wing does not suffer from this issue.

9. References

[1] O Gur and A Rosen, "Comparison between blade-element models of propellers," The Aeronautical Journal, 112, 1138, December 2008, pp. 689-704.

[2] O Gur and A Rosen, "Comparison between blade-element models of propellers," The Aeronautical Journal, 112, 1138, December 2008, pp. 689-704

[3] The Royal Aeronautical Society, "Introduction to installation effects on thrust and drag for propeller-driven aircraft," ESDU 85015, August 2006

[4] H Glauert, Airplane propellers, in Aerodynamic Theory (Vol. 4, Division L), F.W. Durand, (Ed.), New York, Dover, Third Edition, 1935

[5] D Biermann and P E Hartman, "Wind tunnel tests of four and six blade single and dual rotating tractor propellers," NACA Report 747, 1942.

[6] M Drela and M B Giles, "Viscous-inviscid analysis of transonic and low Reynolds number airfoils," AIAA Journal, 25, 10, Oct 1987, pp. 1347–1355.

[7] O Gur and A Rosen, "Propeller performance at low advance ratio," Journal of Aircraft, 42, 2, Mar.-Apr. 2005, pp. 435-441.

[8] C P van Dam, E A Mayda, D D Chao, D E Berg," Computational design and analysis of flatback airfoil wind tunnel experiment," SANDIA report SAND2008-1782, March 2008

[9] G S Lakshmi, P Balmuralidharan, G Sankar, K Selvaraj, and N Balachandran "High lift two-element airfoil design for MALE UAV using CFD," 20th Annual CFD Symposium, 9-10 August 2018, Bangalore

[10] R D Harrington, "Full-scale-tunnel investigation of the static-thrust performance of a coaxial helicopter rotor," NACA TN-2318, March 1951.

[11] K D Korkan, J Camba, and P M Morris, "Aerodynamic data banks for Clark-Y, NACA 4-digit, and NACA 16-series airfoil families", NASA Contract Report No. 176883, January 1986.

[12] D M Dunham, G L Gentry, Jr., and P L Coe, Jr., "Low-speed wind-tunnel tests of single- and counter-rotation propellers," NASA Technical Memorandum 87656, April 1986

[13] D C Mikkelson, B J Blaha, G A Mitchell, and J.E. Wikete, "Design and performance of energy efficient propellers for Mach 0.8 cruise," NASA TM X-73612, 1977.

[14] E Sand, A D Elliott, and V H Borst, "Summary of propeller design procedures and data, Volume III – hub, actuator, and control designs", US Army Air Mobility Research & Development Laboratory Technical Report No. 73-34C, November 1973.

[15] Y Kidron, Y Levy, and M Wasserman, "The EZAir CFD Solvers Suite: EZAir Theoretical and User's Manual - Version 3.54," Tech. Rep. August, ISCFDC, 2016.

[16] Y Mor-Yossef, "Low Mach number enhancement of the AUFSR scheme," Computers and Fluids, 136, 2016, pp. 301–311.

Hydraulic control system of mechanical energy stored type powder high velocity compaction machine

Dongdong You[1,2], Qian Zhu[1], Yanghui Zhu[1] and Fenglei Li[1]

[1] National Engineering Research Center of Near-net-shape Forming for Metallic Materials, South China University of Technology, Guangzhou, 510640, China
[2] Metastable Materials Science and Technology State Key Laboratory, Yanshan University, Qinhuangdao 066004, China

Abstract. An automatic control scheme of spring energy storage by hydraulic control of spring compression is presented and implemented in this paper to improve the automation level, control accuracy and production efficiency of the existing HVC machine with mechanical energy storage type. The hydraulic system, servo system, and PLC codes are developed. The PID control optimization is applied for the control error accuracy of the machine. The transfer function is derived, and the genetic algorithm is used to optimize the PID parameters. The control scheme is implemented in the updated machine. The experimental results of iron powder HVC are in agreement with the predicted and calculated values. The reliability and validity of the automatic control scheme is verified. The research facilitates the industrial application of this kind of equipment.

1. Introduction

High Velocity Compaction (HVC) can effectively improve the densification, product performance and production efficiency of powder compacts. Because of its advantages of high impact speed, low cost, low elastic aftereffect and low demoulding force comparing with traditional pressing methods[1], HVC has been widely used[2-4] and further developed recently[5-7]. The HVC research of iron-, aluminum- and magnetic-based materials indicate that the impact velocity and energy have a critical influence on compact densification during HVC[7-9].

The energy storage methods of HVC equipment include hydraulic, gravity, pneumatic, electromagnetic, and chemical forms[10-13]. Mechanical energy storage machine[14] employs hydraulic pressure to drive mechanical spring, which provides energy and push the hammer to compress powder. Guan[14] carried out the upsetting test of copper pillars on the test machine. The rationality of the machine design was verified by comparing the experimental data. Subsequently, 316L stainless steel powder was tested on a test machine. The maximum compact density of the sample was 7.84 g/cm3, which was higher than 7.336 g/cm3 of the sample produced by the HYP35-7 machine of Nanjing Hilton Precision Machinery Co., Ltd.[15]. In addition, the results of other experiments[7][16], demonstrate that the test machine has good forming ability and can meet the research needs. However, because the impact energy is controlled by manually adjusting the spring compression, the automation level, the control accuracy and the production efficiency of the machine needs to be improved. Then the test machine can't be applied in industrial automation production.

Based on the existing test machine, an automatic control scheme of spring energy storage by hydraulic control of spring compression is presented and implemented in this paper to facilitate industrial application of mechanical energy storage HVC machine.

2. Hydraulic control method

The hydraulic system mainly consists of three hydraulic cylinders, three reversing valves, eight buttons and a Mitsubishi touch screen. The hydraulic servo control frame is shown in Figure 1. After setting parameters on the PLC, the feedback signal and the input signal of the force sensor and the displacement sensor are compared and amplified on the PLC, and then transmitted to the reversing valve to control the opening size of the valve core. Thereby, the amount of hydraulic oil in and out of the cylinder can be adjusted to implement the control of the spring displacement. The spring force signal is read by a pressure sensor installed in the three-way pipe of the inlet of the hollow hydraulic cylinder, and the spring displacement signal is read by a displacement sensor mounted on the hydraulic cylinder. The entire system is a closed-loop control system.

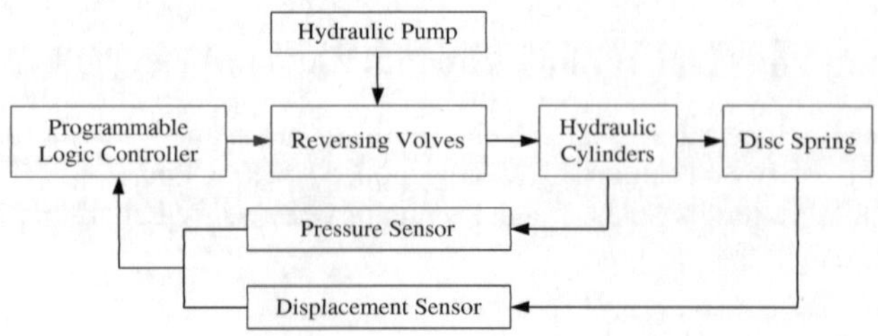

Figure 1. Frame of hydraulic control system

The Mitsubishi FX1N-40MR programmable logic controller with forty control points and GT1150 touch screen are used in the control system. Three reversing valves are driven by 24V built-in current. Both the preset value and the display value can be handled on the touch screen. The preset parameters include stroke, energy, flow, oil pressure upper threshold, start button, and data clear switch. The display parameters include spring force, spring compression, real-time energy value, time required to reach the set energy, real-time flow, and real-time oil pressure.

3. Mathematical model of control system

According to the energy control error of the closed-loop control system, a mathematical model is established to solve the transfer function of the system. The proposed hydraulic control system drives the reversing valve core according to the PLC signal, and then controls the movement of the piston rod of the hollow hydraulic cylinder to realize the automatic control of energy storage. The transfer functions are obtained as follows[17]:

$$G(s)=\frac{K_{sv}A_pK_c}{K_LK_{ec}}\left[\left(\frac{s^2}{w_{sv}^2}+\frac{2\xi_{sv}}{w_{sv}}s+1\right)\left(\frac{s^2}{w_0^2}+1\right)\left(\frac{s^2}{w_0^2}+\frac{2\delta_0}{w_0^2}s+1\right)\right]^{-1} \tag{1}$$

Where w_{sv} is the natural frequency of the reversing valve; ξ_{sv} is the damping ratio of the reversing valve; K_{sv} is the gain; A_p is the effective area of hydraulic cylinder; w_0 is the natural frequency of hydraulic cylinder; δ_0 is the damping ratio; K_{ec} is the flow pressure coefficient; K_L is the gain coefficient. For the displacement sensor, the response speed of the hydraulic cylinder is lower than the response speed of wire encoder pulse, so the encoder link can be regarded as a proportional link, expressed in K_c. The link of the pressure sensor is approximate proportional link for its high frequency bandwidth, expressed in K_f.

The analysis of formula (1) shows that the type 0 system is difficult to meet the fast and quasi-stable characteristics when the signal input. For step function input, the accuracy and rapidity of closed-loop system can be satisfied, but the system may be unstable and oscillate. Hence, the PID module is added to the system to optimize the system control.

3.1. Modeling of PID control for hydraulic system

The AMESim software is used in modeling and simulation. The relationship between the output signal and the input signal is as follow:

$$v = K_p u + K_i \int_{t_{start}}^{t} u dt + K_d \frac{\partial u}{\partial t} \tag{2}$$

Where v is the output signal; u is the input signal; K_p is the scale factor; K_i is the integral coefficient; K_d is the differential coefficient; t is the first order lag time constant.

3.2. Hydraulic control system modeling

The hydraulic cylinder model is created and imported into the mechanical Adams module of spring control, and the integrated model of the hydraulic control system for spring energy storage is established by using AMESim software, as shown in Figure 2.

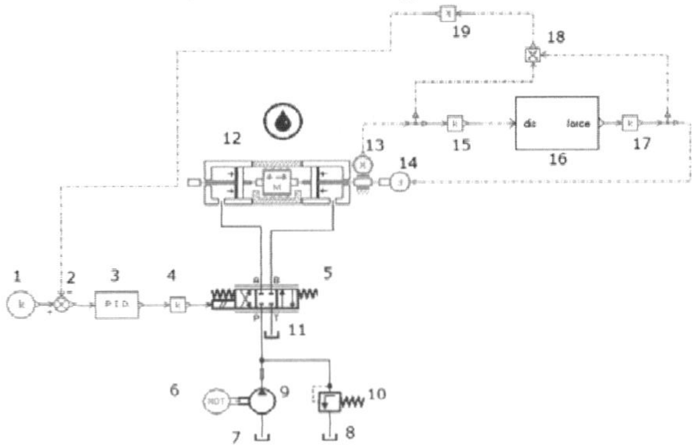

1.control signal; 2.signal comparison; 3.PID control module; 4,15, 17,19.gain; 5.reversing valve; 6-electric motor; 7,8,11.tank; 9.hydraulic pump; 10.overflow valve; 12.hydraulic cylinder; 13.displacement sensor; 14.force sensor; 16.ADAMS module; 18.signal multiplier

Figure 2. Integrated AMESim model of energy control

A series of simulations run one by one in sequence after initializing by different groups of parameters. The genetic algorithm is applied to optimize the PID parameters. The range of PID parameters is determined by AMESim batch processing function. The range of the proportional part K_p is set (10, 20, 50, 100, 200); K_i and K_d are set to 0 and signal 1 is set 1000 J. Then the influence of the parameters variation on the control performance of the system is observed and analyzed.

In the design exploration module of AMESim, the experimental design tool DOE is employed to analyze the parameters and obtain the relationship between the response and the input, and the genetic algorithm in optimization module is used to optimize the PID parameter. Setting the optimized population size 30, the replication probability 0.8, and the mutation probability 0.05, the PID parameters obtained after 36 generations evolution are K_p=105.01, K_i=0.135 and K_d=0.05. Spring deformations calculated using optimized PID parameters are shown in Figure 3.

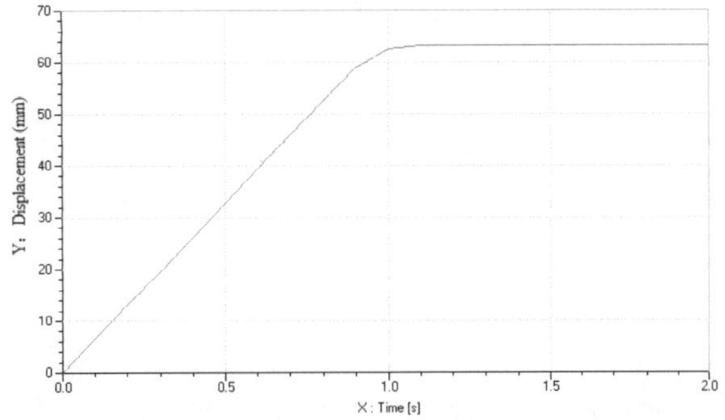

Figure 3. Spring displacement curve after optimization

4. HVC experiment and validation of control system

4.1. Experimental method and device

The experimental material is water atomized iron powder. Two groups of experiments, the energy and the stroke validation tests were carried out. The energy values in each test were set at 200, 500, 800, 1100, 1400, 1700, 2000, 2300, 2600 and 2900 J, and the stroke values in each test were set at 30, 40, 50, 60, 70, 80, 90, 100, 110 and 120 mm, respectively. Two compacts of each specification are produced under the condition of the spring complete releasing. The impact velocity is measured by photoelectric sensor. Then the spring energy storage calculated by the impact velocity is compared with the set energy on touch screen to validate the energy control. The speed measurement results of photoelectric sensor are read by developed Labview software.

4.2. Experimental results

The experimental results of spring energy are compared with the input values and the calculated values, as shown in Figure 4. The measurement results coincide with the set value. The minimum error is 0.7% and the maximum error is 7.2% in the ten sets of tests. The factors causing errors include hydraulic fluctuation, shooting delay, complexity of test environment and limitation of maximum sampling frequency of data acquisition card in the system control. The experimental results indicate that the updating machine has the ability of energy automatic control.

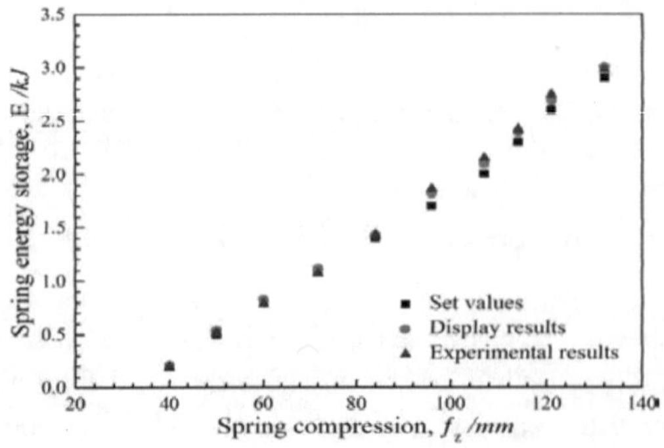

Figure 4. Comparison between set and experiment results

5. Conclusion

(1) A hydraulic control method for spring energy storage is proposed. The hydraulic and servo system are designed with PLC coding. Then the auto control system is implemented by updating the existing machine.

(2) The PID optimization is employed to reduce the control error and promote the accuracy. The optimal solution of the PID parameters is obtained by the genetic algorithm codes.

(3) In both of the energy control and stroke control experiments of the iron powder compaction, the energy storage of the spring is calculated by measuring impact velocity. The experimental results indicate that the proposed control system has the ability of automatic control of stroke and energy. And the relative density of powder compacts reaches a higher value 98.6%. The energy control method is validated.

6. References

[1] Doremus P, Guennec YL, Imbault D and Puente G (2010) *High-velocity compaction and conventional compaction of metallic powders*; comparison of process parameters and green compact properties. Proceedings of The Institution of Mechanical Engineers Part E-Journal of Process Mechanical Engineering **224(3)**:177-185

[2] Daver. *State of the PM industry in North America*[J]. Int J Powder Metall. 2007, **4(43)**: 33-38.

[3] Anon. *A new line in HVC*[J]. Metal Powder Report. 2004, **59(10)**: 5

[4] Andersson. *High velocity compaction of soft magnetic composites netic composites* [J]. Int J Powder Metall. 2006, **40(3)**: 5-11

[5] Hongzhou Zhang, Lin Zhang, Guoqing Dong, Zhiwei Liu, Mingli Qin, Xuanhui Qu and Yuanzhi Lv (2016) *Effects of annealing on high velocity compaction behavior and mechanical properties of iron-base PM alloy*. Powder Technol 288:435-440

[6] Guosheng Jiang, Zhifa Wang, Yi Gu, Qiwang Zhang, Yong Gao and Ken Kuang (2012) *Fabrication of Electronic Packaging Grade Cu–W Materials by High-Temperature and High-Velocity Compaction*. IEEE T Comp Pack Man **2(6)**:1039-1042

[7] Zhao LZ, Deng XX, Yu HY, Guan HJ, Li XQ, Xiao ZY, Liu ZW and Greneche JM (2017) *Structure and performance of anisotropic nanocrystalline Nd-Fe-B magnets fabricated by high-velocity compaction followed by deformation*. J Magn Magn Mater 443:51–57

[8] Yuan XJ, Yin HQ, Din RU, Khan DF and Qu XH (2012) *Study on the impact force and green properties of high-velocity compacted aluminum alloy powder*. Int J Min Met Mater **19(12)**:1107-1113

[9] Li H, Yin HQ, Khan DF, Cao HQ, Abideen Z and Qu XH (2014) *High velocity compaction of 0.9Al2O3/Cu composite powder*. Mater Design 57:546–550

[10] Bos B, Fors C and Larsson T (2006) *Industrial implementation of high velocity compaction for improved properties*. Powder Metall. **49(2)**:107-109

[11] Chen Jin, Xiao Zhiyu, Li Chaojie, Deng Sancai, Leo, Ngai Tungwai and Li, Yuanyuan. *High velocity compaction of 316L stainless powder*. Applied Mechanics and Materials, 2011, 44-47:2993-2997

[12] Sethi G, Myers NS and German RM (2008) *An overview of dynamic compaction in powder metallurgy*. Int Mater Rev **53(4)**:219-234

[13] Meng Zhenghua, Huang Shangyu, Zhou Jing and Sun Wei. *Effect of voltage of powder low-voltage electromagnetic compaction on the density of TiO2 ceramics* [J]. Powder Metallurgy Technology. 2010, **28(1)**: 48-52

[14] Guan HJ, Shao M and Xiao ZY (2012) *Study on mechanical energy stored type powder high velocity compaction machine*. J South China Univ Techno: Nat Sci Ed **40(12)**:53-57

[15] Shao Ming, Guan Hangjian and Xiao Zhiyu. *Experimental Study on Warm Powder High Velocity Compaction Conducted with Mechanical Energy Stored Type High Velocity Compaction Machine* [J]. Journal of Mechanical Engineering. 2013, **49(18)**: 46-51

[16] Chen Jin, Xiao Zhiyu, Tang Cuiyong, Li Chaojie and Zhang Wen. *Investigation on equipment and forming tests of warm powder high velocity compaction*. Materials Science and Engineering of Powder Metallurgy, 2011,**16(4)**:604-609

[17] Maghareh, Amin, Silva, Christian Em Dyke and Shirley J. Servo-hydraulic actuator in controllable canonical form: *Identification and experimental validation. Mechanical Systems and Signal Processing*. 2018,100: 398-414.A. Mecke, I. Lee, J.R. Baker jr., M.M. Banaszak Holl, B.G. Orr, Eur. Phys. J. E 14, 7 (2004)

Acknowledgment

This work was supported by the Natural Science Foundation of Guangdong Province (Grant no. 2017A030313320 and 2015A030312003), the National Natural Science Foundation of China (Grant no. 51875209), and the Open Fund of Metastable Materials Science and Technology State Key Laboratory of Yanshan University (Grant no. 201712).

IOP Conf. Series: Materials Science and Engineering **638** (2019) 012003 doi:10.1088/1757-899X/638/1/012003

A study on decompresses heat pump dryer for drying of shiitake mushrooms at medium temperature

Y J Goh[1], Y H Kim[2], B Lee[2], H C Jung[3], S R Lee[4] and K M Kim[2]

[1]Suncheon Jeil College, Suncheon-si, Jeollanam-do, Republic of Korea
[2]Chonnam National University, Yeosu-si, Jeollanam-do, Republic of Korea
[3]A-ONE Engineering Co.,LTD, Suncheon-si, Jeollanam-do, Republic of Korea
[4]WP Engineering Co.,LTD, Suncheon-si, Jeollanam-do, Republic of Korea

Abstract. This study developed a drying method and a dryer that are optimized for shiitake mushrooms. They are mainly being cultivated and consumed as food and medicine in Korea, Japan, China, and Southeast Asia. They generally contain lots of moisture. For this reason, the use of a dryer causes problems such as decomposition and spoilage, and natural dry processing is preferred. To resolve these problems, various drying methods have been proposed (e.g. hot air drying, heat pump vacuum drying) but there has been a difficulty in maintaining a medium-temperature range of 30 to 40℃, the optimal temperature environments for retaining their main components. The study developed a 2-stage decompresses resistive heat pump dryer that can operate also in medium-temperature environments - existing decompresses heat pump dryers have been able to operate only in high-temperature environments. A dryer optimized for shiitake mushrooms that are being produced and distributed in large quantities was proposed, and the utility of the machine was proved through experiments.

1. Introduction

Shiitake mushrooms are edible mushrooms belonging to Lentinus, Pleurotaceae, Agaricales, Agaricomycetes. They are being cultivated and consumed mainly in Korea, China, Japan, and Southeast Asia. The food of nutritional value that has unique fragrance and flavor contains nutrients (e.g. carbohydrates, protein, minerals, vitamins), and therefore is effectively used as food and medicine[1,2]. Notably, in [3] reported them as having anti-cancer effect and then there was a sharp rise in consumption. They contain around 90% moisture, have the most vitamin C contents among mushrooms, and are considered as an excellent alkaline food.

But because of difficulty in storage (e.g. decomposition, spoilage) caused by their high moisture contents, processing operation methods have been proposed (e.g. bottling, a method where they are salted and then dried by sea breeze)[4]. As for drying, sun-drying has been mostly performed; but there is a problem - it is difficult to obtain products of uniform quality, since they are damaged by spore dispersion and insects and easily spoil due to microbial contamination, etc[5]. To resolve this problem, various methods (e.g. high temperature hot air drying, high temperature heat pump vacuum drying, drying with infrared rays) have been proposed, the methods helping to retain the nutrients and colors for a long period of time when storing them after drying[6].

Incidentally, dryers for agricultural and marine products are important pieces of equipment that facilitate maintaining the long-term freshness of them and storing them for a long period of time by removing the moisture inside them - studies of various drying methods with the use of the machines are being performed and the methods are being widely used in the agriculture and fisheries industry. Dryers can be categorized into grain dryers and dryers for agricultural products, marine products, and forest products according to what is dried; into small and large ones according to capacity; and into ones powered by oil, gas, electricity, hot-air, and heat pump and ones powered by two or more sources (that is, hybrid) according to energy sources for dryers[7].

Hot-air dryers are generally used. They take shorter drying time than sun-drying, but are exposed to high-temperature air for a long time. For this reason, they have these disadvantages - quality deterioration caused by loss of amino acids, surface color change, etc.; high energy consumption while drying[8]. Heat pump vacuum dryers are known to have better energy conservation effect than hot-air dryers, since they can recover the heat used thanks to the nature of their cycle - the difference in energy efficiency between them is more than double[9,10]. Heat pump vacuum dryers are categorized into pressure drying, ambient pressure drying, and low pressure drying according to pressure; low pressure drying is categorized into freeze-drying method and low pressure drying method. In general, freeze-drying method is performed in a high-degree vacuum state and a low-temperature environment. Despite its advantages - the lower degree of damage of food texture; the high retainability of moisture and flavor components - it is mainly used for high-price foods (e.g. coffee, beef) due to high drying costs[11,12].

As for low pressure drying method, it minimizes various chemical reactions, since it is performed at a relatively low temperature, and requires lower drying costs than freeze-drying method. In addition, it produces excellent quality products (e.g. the low oxygen concentration while drying prevents the decomposition and spoilage of the ingredients)[13]. But it has a disadvantage - the high cost of facility and the high operating cost. Also the development of it has not yet progressed, because there are only a limited number of refrigerants that can be used in high-temperature environments.

When drying shiitake mushrooms with a hot-air dryer, the operating temperature range is generally from 40 to 60℃. It is thought that when they are dried at a temperature below 45℃, their quality deteriorates (e.g. spore dispersion; the cap of a shiitake mushroom splits, causing its end to be curled). And it is thought that when they are dried at a high temperature above 55℃, there is concern about them being damaged by the heat, since they initially have high moisture contents[14].

However, a study of shiitake mushroom temperature change and its main components showed that there was no big change in protein contents when the temperature was around 30℃ or higher; the fat contents hardly changed when the temperature was 40℃ or higher; the reducing sugar contents also changed dramatically when the temperature was 30℃; the higher drying temperature meant the lower free amino acid contents; and the higher drying temperature meant the higher amino acids in the protein[15,16]. Judging by these, it can be seen that shiitake mushrooms can remain in optimal condition when the temperature is changed in the medium-temperature range of 30 to 40℃.

Hence the study designed and proposed a decompresses heat pump dryer (DHPD) that operates only under temperature variations, not under wind variations. The machine, optimized for drying them, can dry them at a medium-temperature of 35℃. To prove this, a physicochemical evaluation was performed, and the dryer proved to be excellent by comparative analysis with what was produced by natural drying and hot air drying.

2. Experiment method

2.1. Experimental material

Shiitake mushrooms, the testing materials, were harvested in 2017 at Hansan-myeon, Seocheon-gun, Chungcheongnam-do, Republic of Korea; to minimize their quality change during storage, they were stored in a low temperature refrigerator that worked at 5℃ - 30 kg of raw shiitake mushrooms were used for the drying.

2.2. Experimental conditions

To dry them at a medium-temperature, the optimized environment for drying them, the 2-stage decompresses resistive decompresses heat pump dryer was designed, and they were dried as in figure 1. As for drying methods, sun-drying(natural drying), hot air drying, and the decompresses heat pump were performed. Sun-drying was performed for 3 days (8 hours of sunshine duration per day); hot air drying was performed with the use of Lee Hwa Industry's dryer LH-1200 (the drying temperature: 60 ℃; the drying time: 24 hours); and decompresses heat pump was performed under these conditions - the amount of decompression: -2,000mmAqg; the drying temperature: 35 ℃; the drying time: 18 hours.

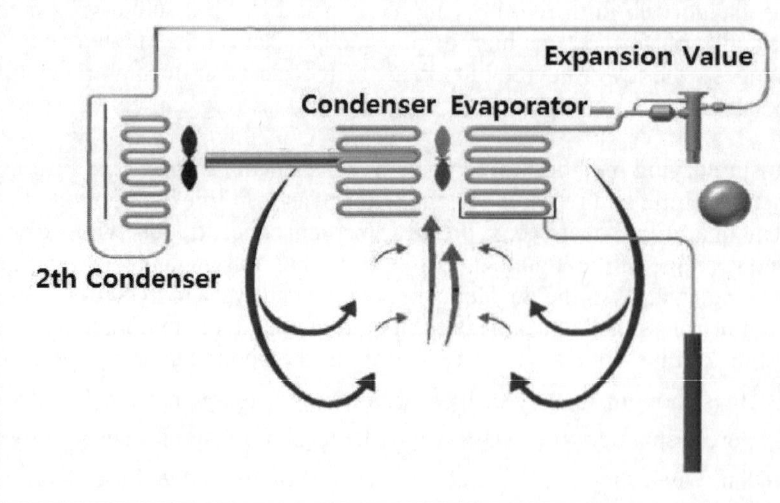

Figure 1. The 2-stage decompresses resistive decompresses heat pump dryer designed

2.3. β-glucan measurement conditions

β-glucan is well documented as lowering plasma lipids in human and animal studies[17-21]. At dietary intake levels of at least 3 g per day, oat fiber β-glucan decreases blood levels of LDL cholesterol and so may reduce the risk of cardiovascular diseases[22,23]. The β-glucan contents were measured by the method in [14] and with the use of Megazyme's Beta-glucan kit. The total glucan value minus the value of α-glucan is the value of β-glucan.

2.4. Total polyphenol content measurement conditions

Polyphenols are abundant micronutrients in our diet, and evidence for their role in the prevention of degenerative diseases such as cancer and cardiovascular diseases is emerging[24].

Over the past 10 y, researchers and food manufacturers have become increasingly interested in polyphenols. The chief reason for this interest is the recognition of the antioxidant properties of polyphenols, their great abundance in our diet, and their probable role in the prevention of various diseases associated with oxidative stress, such as cancer and cardiovascular and neurodegenerative diseases.

100mL of distilled water was added to 5g of the sample. Then it was homogenized and agitated for 15 hours at room temperature. Next, it was centrifuged for 5 minutes at 10,000 rpm, and was filtered through Whatman No. 2 to be constant volume of 100mL. In accordance with Folin-Denis method, an absorbance value was measured at UV spectrophotometer (Infinite200pro, Tecan, Austria) 720 nm; the standard material used was gallicacid (Sigma).

2.5. Antioxidant capacity measurement conditions

As for free radical scavenging activity, the sample reducing power was measured through the method of Brand-Wiliams, etc[17]. 500µg/mL of 0.1mM DPPH in Ethanol Solution was added to 500µg/mL of the extract from the sample. Next, it was mixed well for 10 seconds, and was left as it was for 20 minutes at room temperature, and an absorbance value was measured at 525nm with the use of the microplate reader (Multiscan GO, Thermo Scientific co. ltd., USA). Free radical scavenging activity was expressed as an absorbance ratio (%) between the sample addition group to the non-addition group, and it was calculated through the equation below:

$$\text{Free radical scavenging activity}(\%) = (B-A)/B * 100 \quad (1)$$

Here, A is absorbance value of the sample addition group and B is absorbance value of the non-addition group.

3. Experiment result

3.1. Moisture content

Table 1 shows the moisture contents measured through the loss-on-drying method specified in Korean Food Code. The decompresses heat pump showed the highest efficiency in terms of reduction of moisture contents closely related to food storage and to the maintaining of food quality during distribution - the method reduced the moisture contents by 7.45 % when drying was performed for the same amount of time.

Table 1. Moisture contents of the mushrooms by various drying methods[unit: %].

Drying method	DHPD	Sun drying	Hot air drying	Raw Mushrooms
Moisture contents	7.45	22.23	19.74	84.11

3.2. β-glucan

According to the result of an analysis of β-glucan in each drying sample, the β-glucan contents in the sample produced through the decompresses heat pump were highest (25.94%), those in the sample produced through natural drying were second highest (22%), and those in the sample produced through hot air drying were third highest (21%).

3.3. An analysis of a functional component

According to the result of measurement of the contents of total polyphenol(figure 2), a functional component contributing to antioxidant activation, those were highest in the sample produced through the decompresses heat pump (0.76 µg/g); those in the sample produced through sun-drying and hot air drying and those in raw shiitake mushrooms were similar.

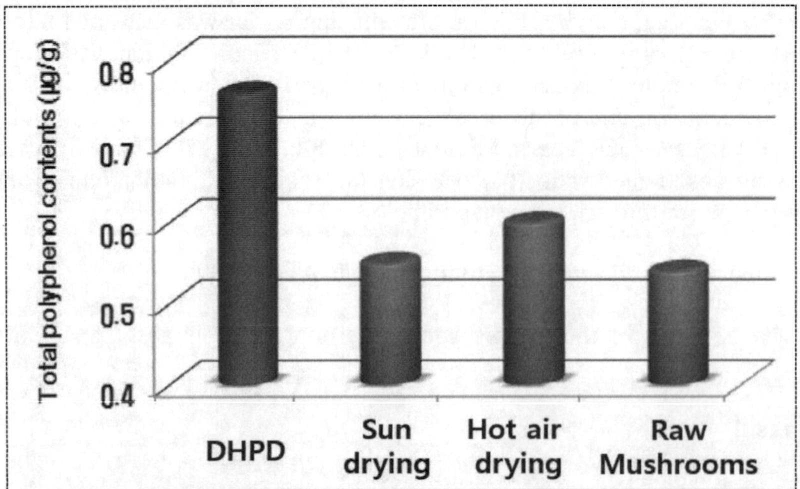

Figure 2. Total polyphenol contents

3.4. An analysis of antioxidant activation

According to the result of an analysis of antioxidant activation through measurement of SOD activation and DPPH radical scavenging of the ethanol extracts from the shiitake mushrooms dried through difference methods, all were concentration-dependent(figure 3). As for SOD activation, at a concentration of 500μg/mL, the decompresses heat pump produced the highest activation (24.0%) - raw shiitake mushrooms: 20.3%; natural drying: 22.7%; hot air drying: 23.3%. As for DPPH radical scavenging, at a concentration of 500μg/mL, raw shiitake mushrooms showed 44.1%, natural drying 35.9%, and hot air drying 40.1%, compared to vitamin C, the control group. The decompresses heat pump produced the highest DPPH radical scavenging activity - 64.4%.

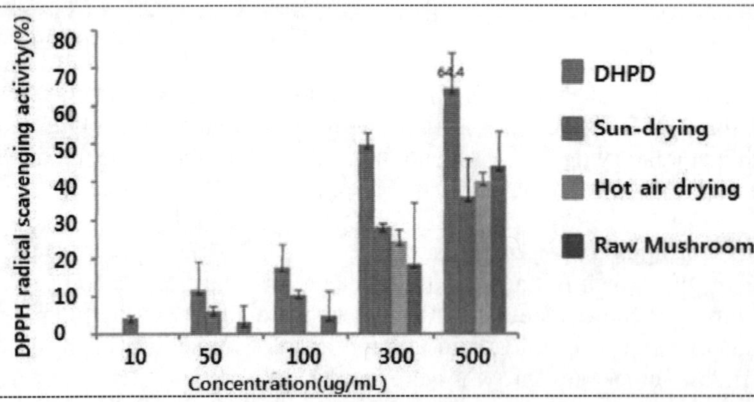

Figure 3. DPPH radical scavenging activity

4. Conclusion

When drying shiitake mushrooms with a hot-air dryer, the operating temperature range is to be from 40 to 60 ℃. When the temperature is below 45 ℃ or is above 55 ℃, problems can arise, including quality deterioration and damage by the heat. Also previous studies even mention these problems of hot-air dryer - failure to dry due to uneven heat transfer; energy efficiency.

Shiitake mushroom component change occurs only in different temperature environments and the temperature condition of 35 ℃ is the optimal condition for preventing that change. In consideration of these, the study designed and proposed the 2-stage decompresses resistive heat pump vacuum dryer that is to operate only in medium-temperature environments (30 to 40 ℃) - with no variations except for temperature. It could be seen that the drying method with the dryer is better in moisture contents, β-glucan, functional component, and antioxidant activation than natural drying and hot air drying performed in these temperature environments. Existing heat pump vacuum dryers operate in high-temperature environments, causing an energy efficiency burden. But it is thought that the proposed method will show good energy efficiency, since it is performed in medium-temperature environments. In the future, the expected energy efficiency aspect needs to be specified, and it is thought that studies of methods of drying various edible and medicinal foods, except shiitake mushrooms, need to be performed.

5. References

[1] J M Sung, Y B Yoo and D R Cha 1998 *Kyohaksa*
[2] S R Han, M J Kim, T J Oh 2015 *J. Kor. Soc. Food Sci. and Nutr.* **44** 1147-49
[3] Y Hiroshi, S Tatsuyuki and H Jyunzo 1987 *J. of Jap. Soc. for cold pre. food* **13** 126-32
[4] C S Han and W S Park 1994 *Doya printing*
[5] V M Choi and J M Soo 1997 *Nat. reas. foun. of kor. report*
[6] Y D Kim, S J Lee, J H Kwon and W G Kim 2015 *J. of Air cond. and refri. eng.* **6** 12-5
[7] C F Jin 2004 *Chungbuk National Univ.*
[8] B Wijesinghe 1997 *CADDET Newsletter*
[9] D B Cho, D P Kim and C S Choi 1981 *J. Kor. Soc. Food Sci. and Nutr.* **10** 53-60
[10] J S Seo, S K Kang and B M Choi 1997 *J. food preserv.* **4** 279-86
[11] K S Ahn C S Kim and J W Lee 2019 *Kor. Soc. Of Sci. and Art* **37** 197-204
[12] H S Yoon, I L Choi, S J Han, J Y Kim, G E Choi, J S Yoon and H M Kang 2017 *J. Kor. Soc. Of Pack. Sci. and Tech.* **23** 83-7
[13] Y J Ji and M H Im 2017 *J. of Food pres.* **24** 60-7
[14] C W Lim, K K Kang, Y B Yoo, B H Kim and S H Bae 2012 *J. Soc. Food Sci. and Nutr.* **41** 1740-6
[15] W B Williams, M E Cuvelier and C Berset 1995 *LWT food sci. Tech.* **28** 25-30
[16] E J Shin and H J Lee 2015 *Trans. Kor. soc. geo. ener. eng.* **11** 22-7
[17] N R Villasmil et al 2007 *J. therap.* **14** 203-12
[18] D Kerckhoffs, G Hornstra and R Mensink 2003 *J. Cli. Nutr.* **78** 221-7
[19] KC Maki et al 2010 *J. ame. Diet asso.* **110** 205-14
[20] K Andersson, K Svedberg, M Lindholmb, R Ostec and P Hellstranda 2010 *J. Athero.* **212** 93-9
[21] I Y Bae, S Y Lee, S M Kim and H G Lee 2009 *J. Food Hyd.* **23** 2016-21
[22] H T Ho et al 2016 *Brit. J. Nutr* **116** 1369-82
[23] J G Lee, K I Kim, I G Hwang, S M Yoo, S G Min and M J Choi 2015 *J. Kor. Soc. Food Sci. and Nutr.* **44** 874-81
[24] C Manach, A Scalbert, C Morand, C Remesy and L Jimenez 2004 *J cli. Nutr.* **79** 727-47

IOP Conf. Series: Materials Science and Engineering **638** (2019) 012004 doi:10.1088/1757-899X/638/1/012004

Ball on the plate balancing control system

Michael M Kopichev, Anton V Putov and Anastasia N Pashenko

Automatic Control Systems Department, Saint Petersburg Electrotechnical University "LETI", ul. Professora Popova 5, 197376, St. Petersburg, Russia

Abstract. The problem of balancing is a continuous study challenge for application in many spheres of interest from education to transportation. Two-degree of freedom ball balancing laboratory test bench is an important plant for control systems understanding. This paper focuses on the mechanical design and control algorithm of balancing a ball on a plate. The plant was equipped with resistive touch screen for ball position data acquisition, stepper motors with position sensors for platform inclination angle feedback data and AVR microcontroller-based two-loop subordinate control system with PID controllers for ball position maintaining. The laboratory bench was designed and assembled in Saint Petersburg Electrotechnical University "LETI" at the Automatic control systems department.

1. Introduction

The problem of an unstable dynamic object balancing is one of the most relevant in the field of of automatic control systems construction [1-11]. Balancing technologies are used nowadays and could be applied in the future in a large number of scientific and technological areas. In addition to demonstrating the effect of the automatic control system and its settings have on a plant, the stabilization problem is used in such spheres as filming, navigation and transport control, including aerial and marine vehicles.

Besides the task of ball on the plate balancing in two-dimensional space (plant with two degrees of freedom shown in figure 1, a), there is a simplified problem of ball and beam balancing (plant with one degree of freedom shown in figure 1, b).

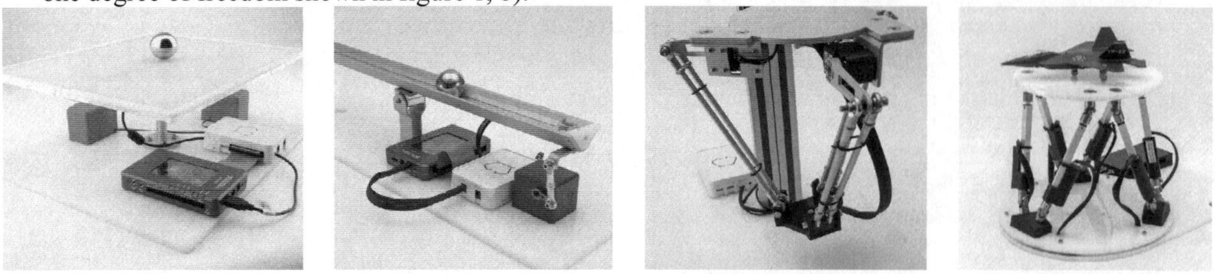

a b c d

Figure 1. Examples of balancing plants.

It is also worth mentioning the task of balancing a dynamic object on a platform was developed in the form of controlling a more complex Delta robots (plant with three degrees of freedom shown in figure 1, c) and the so-called Stuart platform (object with six degrees of freedom shown in figure 1, d), which is the evolution of Delta robots [12].

Ball on the plate, ball and beam and Stuart platform balancing control laboratory test benches could be applied for stabilization of horizontal planes, which is essential for such applications as transportation of fragile objects and stabilization of cameras.

Delta robots are usually used in industrial processes of sorting the product on conveyor lines, so precise flange positioning and stabilization is quite necessary for such kind of robots.
The problem of dynamic object stabilization and control has another application sphere, which so-called ballbots or ball balancing robots. Ballbots are mobile robots that have a single spherical wheel for the implementation of motion, so such design requires continuous self-balancing of the robot while both moving and standing. Due to the single contact spot between ball and surface, the ballbot is able to move freely in any direction and to unfold in one point. Such design increases maneuverability of the robot, but require control system for balancing and moving of the robot. In addition to fueling researchers` interests, ballbots are designed to help older people and disabled people to move and ballbots could probably become a new platform for future transport development. A ball balancing robot that was designed at Saint Petersburg Electrotechnical University "LETI" at the Automatic Control Systems Department that is equipped with three omni-directional wheels with DC motor drives and position sensors, a PIDD control system and MPU-6050 position tracker, as shown in figure 2.

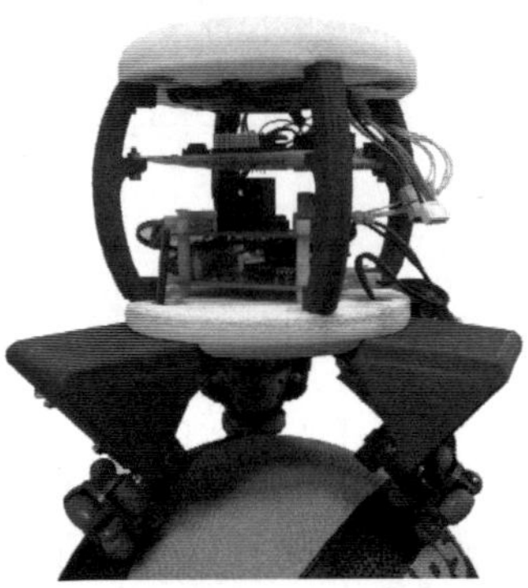

Figure 2. ACS ball balancing robot.

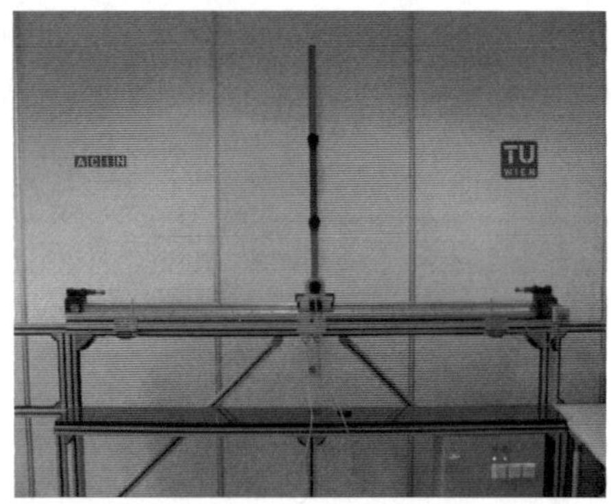

Figure 3. Triple inverted pendulum on a carriage.

In addition to ballbots there is also a segment of inverted pendulums and its application in segways, monowheels and gyroscooters. In 2013 researchers from Austrian university demonstrated the stabilization of the triple inverted pendulum [13]. In figure 3, a triple inverted pendulum laboratory test bench is shown.
In this paper, the problem of dynamic object balancing on a two-dimensional plane is considered, i.e. the plant with two degrees of freedom.
The first step in solving the problem of an unstable object balancing on the plane is tracking current position of this object on a balancing plate. For this purpose, a video camera is often used [2, 6]. The camera is connected to a desktop computer via a standard protocol; it monitors the position of ball using digital filters (if the ball is painted contrast). Typically, Matlab connected to the plant controller is used for control purposes in such cases. For tracking the ball on the plane, a resistive touch screen could be used as well [4, 7]. Touch panel allows user to get rid of the desktop computer with installed Matlab and make the test bench more compact and orderly. Since the use of a video camera, as well as a touch panel, entails costs, and the use of a video camera is also associated with quite a large amount of programming that is not directly related to the problem of balancing, in literature there are non–trivial ways to track the position of the object on the plate, such as the use of a phototransistor grid for instance [3].
The second step in solving the problem of unstable object on plate balancing is the choice of an automatic control system which can provide sufficient positioning accuracy of the dynamic object. The choice of control system is affected by the performance requirements and the set of system components. To solve that problem, as a rule, varieties of PID regulators are used [2-4, 10], however there are also systems with adaptive [8] and fuzzy [9] regulators.
The plant for this work is a laboratory test bench with a platform on which the resistive touch screen is placed. After switching on the test bench, the ball located at some point of the touch screen should be brought to the central point of the sensor. In case of external disturbances (deviations of the ball

position from the center of the touch panel), the object must be returned to the central point. The platform is driven by stepper motors equipped with self-made position sensors based on potentiometers. Two-circuit subordinate control system with P and PD-controllers is designed on the AVR Atmega8A microcontroller.

2. The laboratory test bench mechanical design

A 3D model of the laboratory test bench is shown in figure 4.

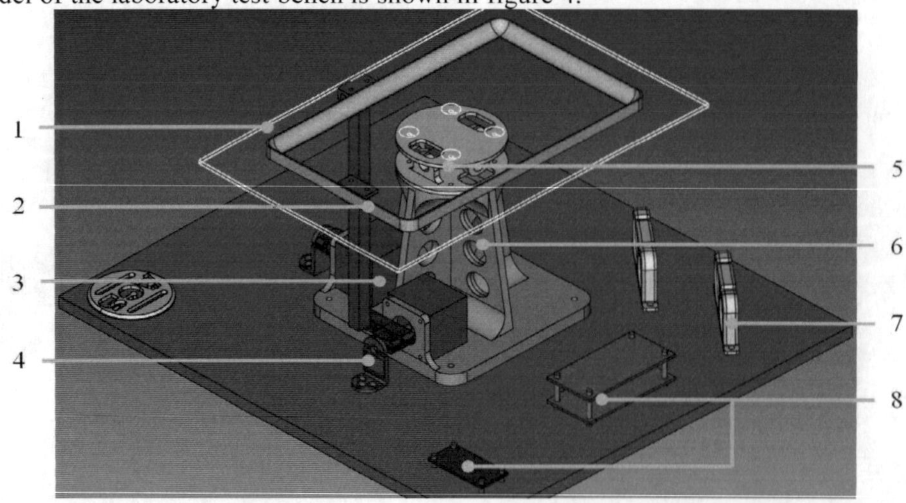

Figure 4. Computer model of the ball balancing test bench.

The laboratory test bench consists of a wooden base on which the carrying "tower" (6 in figure 4) with plexiglas plate, resistive touch screen (1 in figure 4), stepper motors (3 in figure 4) with rods used for changing an angle of the platform inclination, motors` shaft position sensors (4 in figure 4) for platform inclination feedback, control boards with motor control drivers and microcontroller and power supply unit (8 in figure 4) are placed. A plate with a resistive touch screen is mounted on the "tower" with a gimbal suspension (5 in figure 4) to limit the degrees of freedom of the platform to two and equips with a restrictive frame on the edge of the platform (2 in figure 4) to prevent the impossibility of reading the position of the ball on the sensor. Moreover, the wooden base carries metal balls` of different weigh (7 in figure 4) cradles for their easy storage and carrying. The platform "tower" with motors` mounts, gimbals, restrictive frame, metal balls cradles and protective elements of the laboratory test bench were printed on a 3D printer of ABS plastic. The size of the item is 400×400×220 mm.

To determine the position of the ball on the plate, a four-wire 9-inch touch screen was used. The touch panel is connected to two channels of the ADC of the microcontroller to determine the X (longitudinal) and Y (transverse) coordinates of the object. The advantages of using the touch screen includes the possibility to work with the microcontroller directly and as a result short response time, as well as its cost. The disadvantages includes the presence of noise and, as a consequence, the need to filter the sensor`s signal.

Stepper motors with A4988 chip driver and a potentiometer for providing the motor`s shaft position feedback were chosen for motion implementation. Stepper motors are connected directly to the driver, which in turn is connected to the microcontroller by only two wires (STEP and DIRECTION). Drivers of the stepper motors are set to half-step operation mode. Motors potentiometers of the longitudinal and transverse axis of the platform displacement are connected to two more channels of ADC of the microcontroller. The advantages of using stepper motors include high torque and greater reliability. The disadvantages includes heating from long working time and necessity to install a driver for providing micro-step control modes. To reduce the heat dissipation, the stand is equipped with a standalone motors shutdown button.

The laboratory test bench is shown in figures 5 a, b.

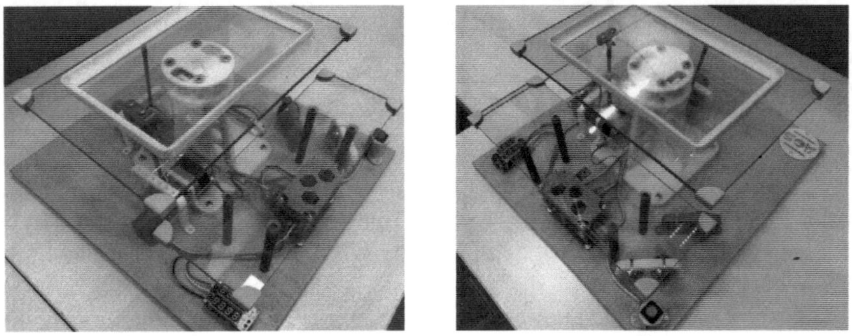

a b

Figure 5. Ball balancing laboratory test bench.

For the data acquisition and processing as for generating the control action, the AVR microcontroller Atmega8A was chosen. The control algorithm of the test bench is written in C programming language in the development environment CodeVisionAVR. For the convenience of program debugging, the wireless UART data transfer protocol was used for information exchange between microcontroller and computer.

3. Ball balancing control system

A resistive touch panel is used to determine the position of the object on the table. Angle of inclination of the table is determined by potentiometers that track the angle of rotation of the stepper motors` shafts. The control system operates at a frequency of 100 Hz and involves reading information about the coordinates of the ball and the angle of inclination of the platform. The basis was adopted by a two-circuit subordinate automatic control system to maintain the position of the ball on the platform with versions of PID controller. The outer contour is responsible for controlling the position of the ball (data obtained from the resistive screen), and the inner – for the position of the platform (data obtained from the motor position sensors).

In figure 6, the structure of the designed automatic control system is shown.

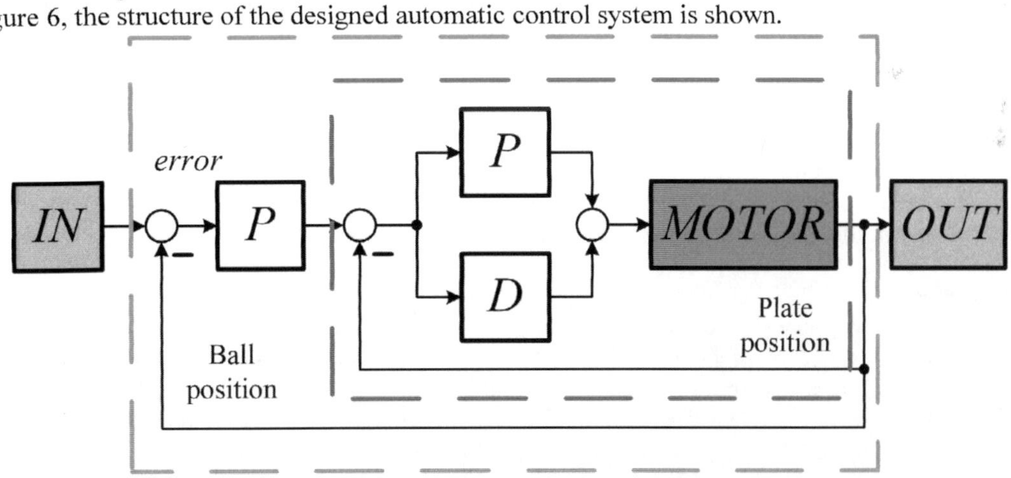

Figure 6. Ball on the plate balancing control system diagram.

The goal of the control system is to bring the object into the centre point of the resistive touch screen and hold it in that point despite all the disturbances by changing the inclination angle of the plate.

4. Control system test

Figures 7 a, b shows the qualitative dependence of time to the coordinates (number 1 indicates the longitudinal coordinate X of the ball change, number 2 – transverse coordinate Y change) of the 170 grams steel ball, obtained by wireless channel UART. At the initial time the ball was placed in the corner of the table, then the motors were switched on and control systems started working. Two plots were obtained for different disturbances.

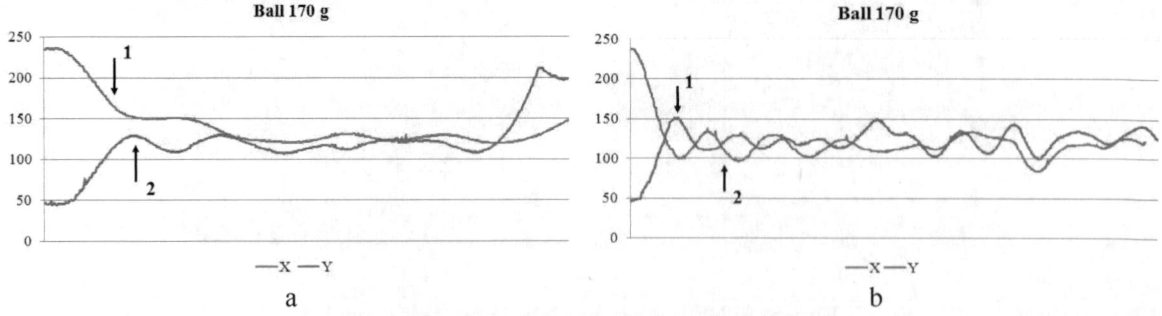

Figure 7. Ball balancing experimental results.

Due to half-step operation mode of the stepper motors the position of the ball should always be corrected by the control system. The step response settling time is about 3 seconds. Resistive touch screen ADC signal was filtered for noise cancelling. Based on the data obtained, it can be stated that control system manage with the task of sustaining the position of the ball at the central point of resistive touch screen.

5. Conclusions

In this paper the design of a ball balancing laboratory test bench was described. The object designed could be used for testing the stabilization and balancing control algorithms, which could be essential for instance for marine ships and aerial vehicles. The microcontroller-based control system tracks the ball position on the resistive touch screen and calculates the position error, which is the displacement of the coordinates of the ball relative to the central point of the resistive glass. Stepper motors with a position sensor allows control system to change the inclination angle of the platform with resistive glass for moving the ball on the table. Microcontroller two-loop subordinate control system with varieties of PID controllers allow to maintain the ball in the centre point of the resistive panel. The results obtained are important due to use of stepper motors instead of usual servos and lack of any external computer for commonly used computer vision system. Instead of it the described version of laboratory test bench could be programmed directly from user`s computer because of microcontroller based control system availability. In the future, it is planned to improve the filtering of the ball position feedback signal to increase the speed and quality of the automatic control system. Also, described problem could be transformed into more complicated delta robot control that could be used in many spheres of industrial automation from printing and milling to assembly lines.

6. References

[1] Knuplez A, Chwdhury A and Svecko R 2003 Proc. of the IEEE international conf. on industrial technology, *Modelling and control design for the ball and plate system*, p. 1064-1067

[2] Jadlovska A, Jajcisin S and Lonscak R 2009 Proc. of the 17th int. conf. on process control, *Modelling and PID control design of nonlinear educational model ball & plate*, p.475-483

[3] Zeeshan A, Nauman N and Jowad Khan M 2012 Proc. of the 9th international Bhurban conference on applied science & technology, *Design, control and implementation of a ball on plate balancing system*, p. 22-26

[4] Koszewnik A, Troc K and Słowik M 2015 Acta Mechanica et Automatica, *PID controllers design applied to positioning of ball on the Stewart platform*, **8(4)**, p. 214-218

[5] Gao Z, Wijesinghe S, Pathinathanpillai T, Dyer E and Singh I 2015 Int. Journal of Mechatronics, Electrical and Computer Technology, *Design and implementation a ball balancing system for control theory course*, **5(17)**, p. 2363-2374

[6] Debono D and Bugeja M 2015 Proc. of the 12th int. conf. on informatics in control, automation and robotics, *Application of sliding mode control to the ball and plate problem*, p. 412-419

[7] Chan Ham C 2015 Proc. of the 122nd ASEE annual conf. & exposition, *Development of a ball-on-plate system*
[8] Xiao J and Buttazzo G 2016 Proc. of the 8th int. conf. on adaptive and self-adaptive systems and application, *Adaptive embedded control for a ball and plate system*, p. 40-45
[9] Kuncan M, Kaplan K, Acar F, Kundakci I and Ertunc H 2016 Int. Journal of Natural and Engineering Sciences, *Fuzzy logic based ball on plate balancing system real time control by image processing*, **10(3)**, p. 28-32
[10] Shih C L, Hsu J H and Chang C J 2017 Journal of Computer and Communications, *Visual feedback balance control of a robot manipulator and ball beam system*, **5**, p. 8-18
[11] Khan K A and. Ryu J C 2017 Proc. of the 2017 ASEE annual conf. & exposition, *ROS-based control of a manipulator arm for balancing a ball on a plate*
[12] https://www.acrome.net
[13] Glück T, Eder A and Kugi A 2013 Automatica, *Swing-up control of a triple pendulum on a cart with experimental validation*, **49 (3)**, p. 801-808

Research on the friction pair of the space high-power and long-life rolling ring

Chengshan Liu [1], **Xiuqun Chen** [1] **and Zhiyuan Qian** [1,2]
[1] Aerospace System Engineering Shanghai, Shanghai 201109, China
[2] Shanghai Key Laboratory of Spacecraft Mechanism, Shanghai 201108, China

Abstract. With the development of spacecraft, the requirement for high-power, long-life and high-reliability electrical transmission technology of the rolling ring is increasingly urgent. Aiming at the friction pair's structure of the rolling ring, relevant theoretical and experimental research was carried out. The kinematic model is established to determine the friction pair's structure of the rolling ring, and the zero sliding of friction pair is verified by ADAMS simulation. The electrical contact resistance and fatigue life of the friction pair are theoretically analyzed, and the cycling life of the flexure was analyzed. The stress distribution of flexure during rolling was simulated by FEM. The lifetime experiment of 1.5×10^5 r that the prototype transmitted the 6.6 kW of single loop was verified. The experimental result shows that the electrical contact resistance and the fluctuation value of the single loop are less than $1\text{m}\Omega$ during the rolling, although the transmission power increased to 10 times as compared with the traditional slip ring. The kinematic model can provide a certain theoretical reference for design of the space high-power and long-life rolling ring.

1. Introduction

The collector ring is a device for electrical transmission between two continuous relative rotated equipment, which generally adopt the slip ring or the rolling ring [1]. The collector ring is widely used in various solar array drive mechanism (SADM), scanning drive mechanism (SDM) and other drive mechanisms of spacecraft whose operational reliability is directly related to the energy supply of spacecraft, that is, the collector ring is the life line of the spacecraft [2]. With the rapid development of aerospace engineering technology, the spacecraft has the long-time requirement of more than 15 years on low earth orbit (LEO), high-power and high-reliability electrical transmission. In contrast to the traditional slip ring, the rolling ring substitutes the rolling friction pair for the sliding friction pair of the traditional slip ring, which reduces the friction and wear of the friction pair and remarkably improves the operating life, transmission power and the electrical contact property (ECP) [3]. Hence, the rolling ring become a preferred solution to achieve the high-power and long-life electrical transmission for SADM [4]. Rolling ring had applied for spacecraft after several decades of development since its birth [5]. Power supplies of the rolling ring in international space station (ISS) has attained 65.5kW, which consists 24 power loops. To study the electrical transmission characteristics of the rolling ring, the various experimental studies on rolling ring applied for ISS has been performed by NASA [6-7]. Zheng et al. [8] built a model to obtain proper operating stress and structure parameters in flexure design to ensure its reliable electrical contact between the inner and outer conducting ring. Hou et al [9] proposed the primary schematic of an innovative Multi-Rotary joints concept by using various power rolling ring to solve the single point failure of the space station. Liu et al [10] established the model of contact resistance to investigate the rolling ring's key parameters and finished the steady-state thermal analysis of different current transferring.

However, the above theory of the rolling friction pair's design is still not perfect and requires to be improved. The objective of this work was to establish the kinematic model of the rolling friction and verified its reasonability. Verification was conducted by means of both numerical simulation (FEM and ADAMS) and experimental tests. The motivation for this work was to produce a general model, obtained the key structural parameters achieving the rolling ring's high-power and long-life electrical transmission for SADM, and provided a theoretical reference for the design of the rolling ring.

2. Structural design and analysis

2.1. Friction pair structure design

In this paper, the space high-power rolling ring is composed of the inner and outer conducting rings, flexures, idlers and other parts and its configuration is shown in Fig. 1. The rolling friction pair is the core component of the rolling ring, including the inner and outer conducting rings, the flexure, the idler and the guide rail. Multiple pre-deformation flexures are embedded between the inner and outer conducting rings, which is used to obtain a greater transfer current capacity. The outer conducting ring fixed, the inner conducting ring rotates clockwise, and multiple flexures rotate counterclockwise and revolve clockwise, which acquires high-power electrical transmission between the inner and outer conducting rings.

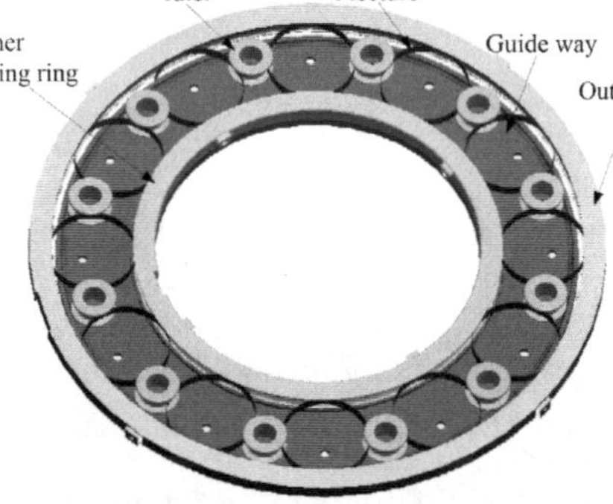

Figure 1. Structure of the high-power rolling ring

To avoid sliding friction at rolling interface, the structural size of various rolling components must be ensured to realize the zero sliding of the friction pair. The inner edge of idler is used to provides rolling interface at every flexure and idler contact interface, which prohibit sliding between adjacent flexure. During the rolling, the outer edges of the idlers are piloted by a guide rail that is fixed to and rotate with the inner conducting ring, which averts motion interference between the flexures or influence to electrical transmission performance of the rolling friction pair. The schematic of the rolling friction pair is shown in Fig. 2, and the zero sliding analysis is performed in the next.

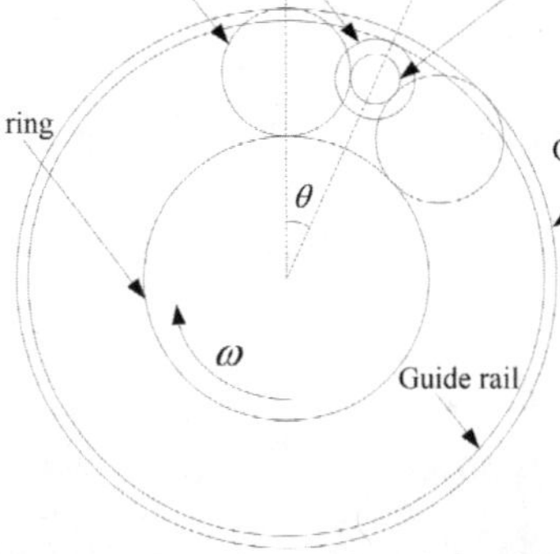

Figure 2. Schematic of the friction pair

The simplified schematic of kinematic model is shown in Fig. 3. There are two basic drive chains in the rolling friction pair from Fig. 3(a), which respectively are recorded as L_1 (inner conducting ring—flexure—outer conducting ring) and L_2 (OO_A—idler—guide rail).

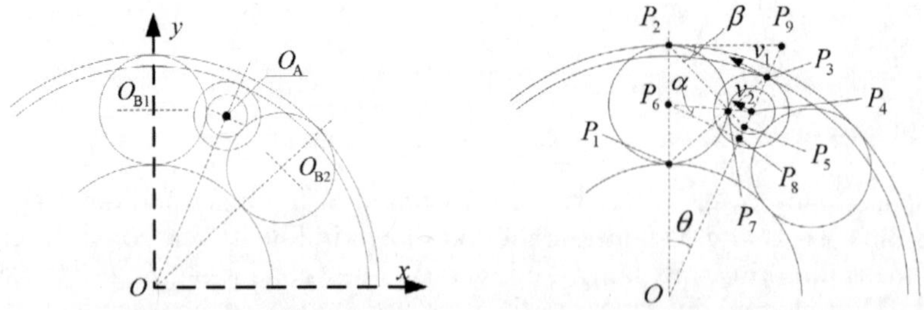

(a) Diagram of the drive chain (b) Schematic of the kinematic model

Figure 3. The kinematic model of rolling friction pair

Assuming that the various elements are zero sliding at the rolling interface, the angular velocity's relationship for two drive chains expressed as

$$\begin{cases} L_1 : \begin{cases} \dfrac{\omega_1 - \omega_H}{-\omega_H} = -\dfrac{R_O}{R_I} \\ \dfrac{\omega_1 - \omega_H}{\omega_1} = -\dfrac{r}{R_I} \end{cases} \\ L_2 : \dfrac{\omega_3}{\omega_1 - \omega_H} = \dfrac{R_B}{R_A} \end{cases} \tag{1}$$

Where R_I and R_O respectively represent the radius of the inner and outer conducting ring; R_B represents the radius of the guide rail; r represents the radius of the flexure; R_R represents the radius of the inner edge of the idler; R_A represents the radius of the outer edge of the idler; ω_1 represents the angular velocity of the inner conductive ring and the guide rail; ω_H represents the revolution angular velocity of the idler; ω_2 represents the rotation angular velocity of the flexure; ω_3 represents the rotation angular velocity of the idler.

Since the drive chains L_1 and L_2 have an equivalent angular velocity, the velocity matching condition at their engagement point is given as

$$R_R\omega_3 = -r\omega_2 \tag{2}$$

The instantaneous centers and velocity vectors involved of every element are shown in Fig. 3(b). The velocity vectors of P_3 and P_4 are respectively recorded as $\mathbf{v}_1$ and $\mathbf{v}_2$. Because the instantaneous center of the flexure is P_2, the velocity direction at P_7 of the flexure is along the line P_7P_1. Therefore, the velocity direction of the idler at P_7 must also follow the line P_7P_1. According to the velocities of the idler at P_7 and P_3, the instantaneous center of the idler is P_5 and the following relationship is obtained, expressed as

$$\begin{cases} \dfrac{v_2}{v_1} = \dfrac{\overline{P_5P_4}}{\overline{P_5P_3}} = \dfrac{\overline{OP_4}\omega_H}{R_B\omega_1} \\ \overline{OP_4} = R_B - R_A \end{cases} \tag{3}$$

Considering the geometric relationship in Fig. 3(b), the α and β are deduced

$$\begin{cases} \beta = 45^\circ + \dfrac{\theta - \alpha}{2} \\ \alpha = \arccos\dfrac{(R_I + r)\sin\theta}{r + R_R} \end{cases} \tag{4}$$

Where α is the acute angle between the lines $\overline{P_4P_6}$ and $\overline{P_6P_8}$; β is the acute angle between the lines $\overline{P_2P_5}$ and $\overline{P_2P_9}$; θ is the acute angle between the lines $\overline{OP_2}$ and $\overline{OP_5}$.

Similarly, the analytical expressions of lines $\overline{P_2P_5}$, $\overline{OP_5}$ and $\overline{P_4P_6}$ are given as

$$\begin{cases} \overline{P_2P_5}: & y-(R_{\mathrm{I}}+2r)=-\tan\beta x \\ \overline{OP_5}: & y=\cot\theta x \\ \overline{P_4P_6}: & y=-\tan(90^{\circ}+\alpha)x+(R_{\mathrm{I}}+r) \end{cases} \tag{5}$$

According to Eqs. (1), (2), (4) and (5), the x and y coordinates of P_4 and P_5 can be obtained, which is substituted into Eq. (6) and leads to

$$\begin{cases} \dfrac{R_{\mathrm{I}}+r}{\cot\theta-\cot(90^{\circ}-\alpha+\theta)}=(R_{\mathrm{B}}-R_{\mathrm{A}})\sin\theta \\ R_{\mathrm{A}}-R_{\mathrm{B}}=-\dfrac{2(R_{\mathrm{I}}+r)^2R_{\mathrm{B}}}{(R_{\mathrm{I}}+r)R_{\mathrm{I}}+R_{\mathrm{B}}(R_{\mathrm{I}}+2r)\left[\tan(\dfrac{90^{\circ}-\alpha+\theta}{2})+\cos\theta\right]} \end{cases} \tag{6}$$

The Eq. (6) implies the variables R_{A} and R_{B} . Given that the design parameters r and R_{I} are confirmed, it can be considered to solve Eq. (6) by the Quasi-Newton Methods.

The size design of the rolling friction pair that is composed of the inner and outer conducting rings, the flexures, the idlers and the guide rail is completed according to the above analysis. The size design ensures zero sliding at rolling interface of the rolling friction pair, which minimizes the sliding friction to reduce wear debris.

In space high-power rolling ring, contact resistance and fatigue life are the two most important indicators. The two parameters are analyzed in the next.

2.2. Fatigue life

During the rolling of the flexure, the instantaneous force analysis is shown in Fig. 4. Assuming that the flexure remains circular during the rolling, it is according to the analysis of the statically indeterminate structure in mechanics of material that the bending moment and the maximum stress of the flexure's arbitrary section are obtained, which is given by

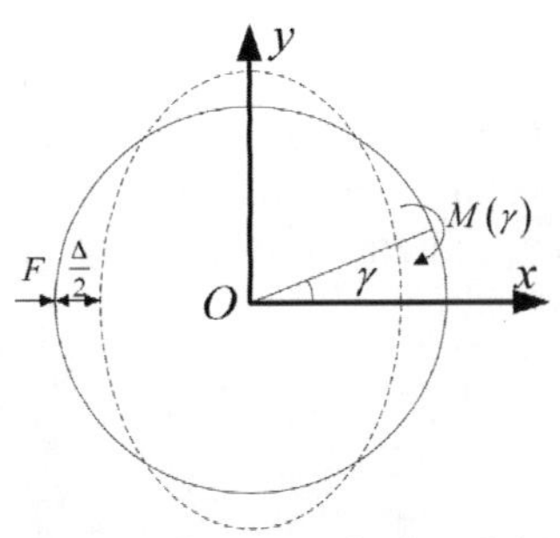

Figure 4. The force analysis of the flexure

$$M(\gamma)=Fr\left(\frac{1}{\pi}-\frac{\cos\gamma}{2}\right) \tag{7}$$

$$\sigma_{\max}=\frac{Mt}{I} \tag{8}$$

Where γ denotes the angle between a section of the flexure and the positive direction of the x -axis.

Based on the material mechanics, the relationship between the deformation of the flexure and the radial bending force is given as

$$\Delta=\frac{Fr^3}{EI}\left(\frac{\pi}{4}-\frac{2}{\pi}\right) \tag{9}$$

According to Eqs. (7), (8) and (9), the maximum stress of arbitrary section is obtained as

$$\sigma_{\max}(\gamma)=2\frac{Et\Delta}{r^2(\pi^2-8)}(2-\pi\cos\gamma) \tag{10}$$

During the clockwise rotation of the flexure, the bending moment and the maximum stress of section change periodically. When the flexure rotates 0°, the stress reaches the minimum; when the flexure rotates 90°, the stress reaches the maximum. According to Eq. (10), the theoretical maximum stress of the flexure is 87.1 MPa in this design.

According to the S-N curve of the QBe 2.0 [11], the fatigue life of the flexure was analyzed, as shown in Fig. 5.

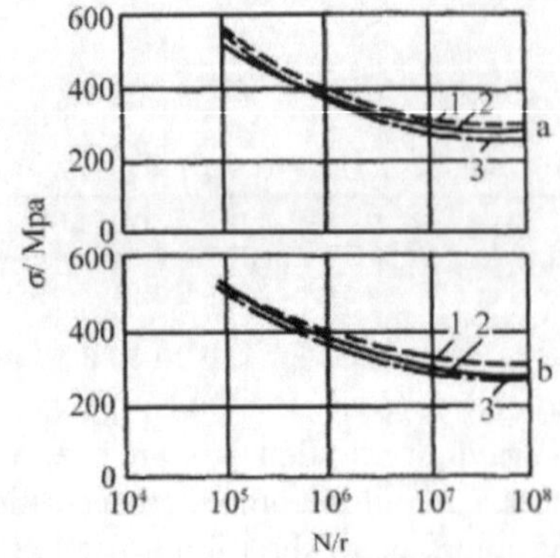

Figure 5. The fatigue life of beryllium bronze

During the rolling of the flexure, the maximum of theoretical stress is 87.1 MPa, which is less than the corresponding stress of 10^8 r for QBe2.0 (>200 MPa). Therefore, according to the theoretical results, the flexure can achieve the lifetime of the 10^8 r, which meets the requirements of the long-life electrical transmission of spacecraft.

3. Simulation analysis

3.1. Zero sliding analysis

The rolling friction pair was designed by the establishment of the kinematic model. It is using the ADAMS that analyzes the linear velocity of every element at the rolling interface to verify the zero sliding of designed friction pair.

At the point of the flexure contacting with idler, the linear velocities of the flexure and the idler was comparatively simulated. The result is shown in Fig. 6. It can be seen in Fig. 6 that the velocity ratio of the flexure to the idler is approximately 1 at the contacting point, that is, the flexure and the idler are zero sliding at the rolling interface.

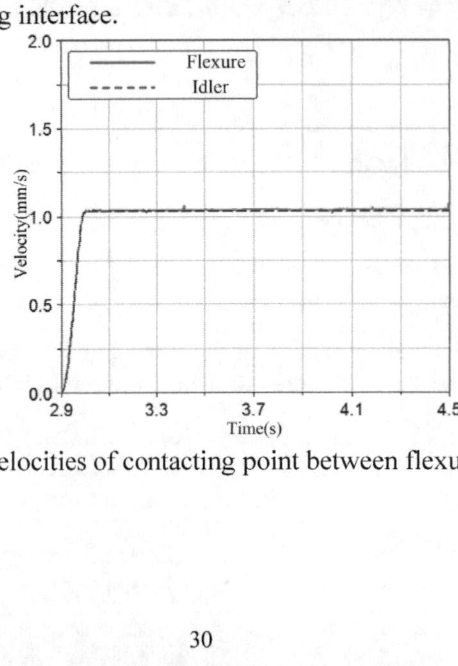

Figure 6. The velocities of contacting point between flexure and idler

At the contacting point between the flexure and the fixed outer conducting ring, the linear velocity of the flexure was analyzed, which is shown in Fig. 7. It can be obtained from the simulation analysis that the velocity of the flexure is 0 at the rolling interface, and the velocity of the outer conducting ring is 0. Therefore, it is not possible that the flexure and the outer conducting ring slide at the contacting point. In other words, the flexure and the outer conducting ring are zero sliding at the rolling interface.

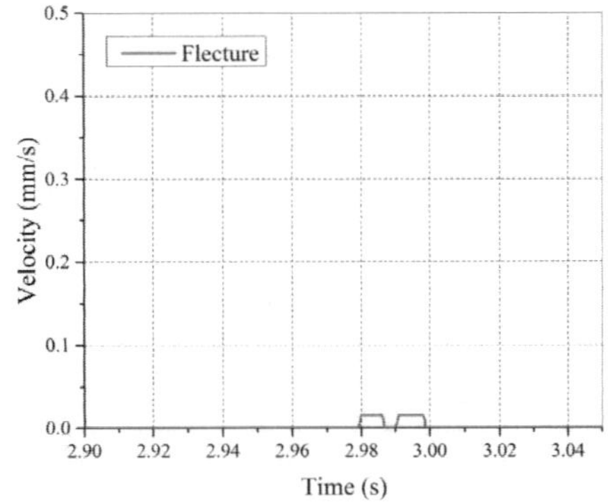

Figure 7. The velocity of contacting point between flexure and outer ring

According to the simulation analysis of the velocity, the flexure is zero sliding at the point of contacting with the idler and the outer conducting ring. Hence, it can be deduced that the flexure is impossibly sliding at the rolling interface with the inner conducting ring from theoretical analysis based on the simulation results in above.

3.2. Fatigue life

During the rolling of the flexure, the maximum stress of the flexure was verified by FEM simulation. Since each flexure has the same forced state, it can be simplified as a single flexure to cooperate with the inner and outer conducting rings. The small displacement collision model was used to analyze the stress distribution of the flexure in simulation at the rolling interface, the inner and outer ring. The material properties during the FEM simulation of cyclic stress are shown in Table 1.

Table 1. Material properties of fatigue life analysis

Desciption	Marerial	Density(kg/m^3)	Young's modulus (GPa)
Inner and outer ring	Copper	8900	123.5
Flexure	QBe 2.0	8300	137.0

According to the asymmetric deformation of the flexure, the stress exists a slight difference at the area where the flexure respectively contacts with the inner and outer ring. The overall stress distribution of the flexure is shown in Fig. 8. From Fig. 8, the stress near the outer conducting ring is slightly larger with the maximum of 82.46 MPa. The stress near the inner conducting ring is slightly smaller, and the maximum is 81.98 MPa.

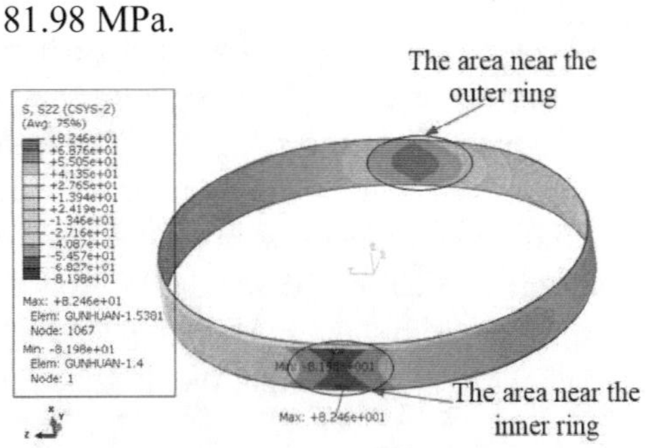

Figure. 8 The stress distribution of the flexure

Considering the results of the theoretical calculation and simulation, the maximum stress of the flexure is less than 87.1 MPa during the rolling. When the cycle life of the QBe 2.0 reach 10^8 r, the stress of the material is larger 200MPa. Hence, based on the analysis of theory and FEM simulation, the flexure could achieve the lifetime of 10^8 r.

The lifetime of the rolling ring refers to the rotations of the inner conducting ring. According to the Eqs. (1) and (2), the relationship of the angular velocity between the flexure and the inner ring can be obtained. When the flexure rotates 10^8 r, the rotations of the inner conducting ring is 1.4×10^7 r, which is the lifetime of the rolling ring.

The speed of the SADM is less than 0.1 revolutions per minute (rpm) on LEO. Hence, the rotations of the SADM is less than 10^4 r during one year. The long-life requirement of the SADM is generally 15 years on LEO, that is, the lifetime of the SADM is required to be 1.5×10^5 r. According to the above analysis of the theory and FEM simulation, the lifetime of the designed friction pair of the rolling ring is much longer than 1.5×10^5 r, which can meet the long-life electrical transmission requirements of the SADM.

4. Experimental verification

On the basis of theoretical calculation and simulation analysis of the friction pair, a space high-power rolling ring prototype was developed, as shown in Fig. 9. The prototype consists of four loops that the assembly shown in Fig. 1 is a loop. Each loop is applied with the nominal 60A current and 110V voltage and the speed of inner ring is 20 rpm, which is used to verify the transmission capacity and the ECP of the prototype.

Figure 9. The prototype of high-power rolling ring

The contact resistance is composed of the constriction resistance and the film resistance. Since the area of contact spots is smaller than the nominal contact area, the current is shrunk when the current is transmitted through the " α spot" between the contact interfaces, so that the resistance is increased. This resistance is constriction resistance. In addition, the contaminated film of the contact surfaces increases the resistance of the contact spot, and the resistance is the film resistance. The factors affecting the contact resistance are complex, such as contact material, contact form, surface roughness, contact load, surface film state, etc [12-13].

As shown in Fig. 10, the ECP of each loop was tested in the experiment campaign whose the contact resistance represent a increasing trend during the rolling of the prototype. Due to the rising of the temperature and the wear in the power track, the contact resistance approximately increased from $0.1\text{m}\Omega$ in average to $0.8\text{m}\Omega$ in average in the vacuum environment. Each loop can transmit the power of 6.6kW, which meet the high-power transmission requirement of the SADM on LEO. When the lifetime of the prototype reached 1.5×10^5 r, the maximum and the fluctuation of the contact resistance were less than $1\text{m}\Omega$, which presents the excellent electrical performance during the test. Hence, though the qualification of test, it was proved that the designed prototype is able to realize the long-life electrical transmission.

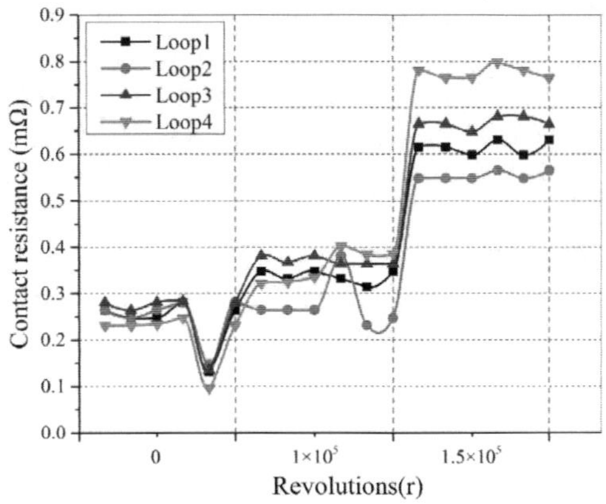

Figure 10. The contact resistance in prototype at 60A during the rolling

Through the experimental verification, the prototype of the rolling ring was proved to be able to achieve the high-power and long-life electric transmission and was robust during the long-life test of 1.5×10^5 r.

5. Conclusion

In this paper, an innovative analytical kinematic model is proposed to provide a theoretical reference for the friction pair's design of the space high-power and long-life rolling ring for the SADM on LEO. The following conclusions are drawn:

(1) Based on the designed parameters, the prototype of the rolling ring has been proved to have the excellent ECP that the contact resistance is less than $0.8\text{m}\Omega$ during the experiment verification of 1.5×10^5 r.

(2) In the test, transmitting the 6.6kW, the single loop of the prototype presents remarkable improvement at power transmission, which is approximately 10 times larger than the traditional slip ring.

(3) The experimental result shows that the design can meet the high-power and long-life requirement of the electrical transmission for the SADM. The research results can provide a theoretical reference for the engineering development of the high-power and long-life rolling ring.

6. Reference

[1] Liu Zili, Hou Xinbin, Wang Li, et al. Thermal analysis of novel space high power rolling ring[J]. Spacecraft Engineering, 2017, **26**(4): 52-59. (in Chinese)
[2] Space International. The development of Chinese manned spacecraft[J]. Space International, 2013(7): 20-26. (in Chinese)
[3] Liu Wenke, Zheng Chuanrong, Zhao Kejun. Overview on development and application of roll ring technology[J]. Fire Control Radar Technology, 2014(2): 107-111. (in Chinese)
[4] J. Batista, J. Vise, K. Young. Roll Ring Assembles for the Space Station[R]. Washington DC: National Aeronautics and Space Administration, 1994.
[5] National Research Council. Laying foundation for Space Solar Power: An Assessment of NASA's Space Solar Power [M]. Washington, D.C., National Academy Press, 2001.
[6] Kasper R. Electric power from orbit: A critique of a satellite power system[J]. NASA STI/Recon Technical Report N, 1981, 82.
[7] Renz D D. Multi-hundred kilowatt roll ring assembly evaluation results[C]. Intersociety Energy Conversion Engineering Conference. 1988.
[8] Zheng Chuanrong, Zhao Kejun. Study on flexible ring design of roll ring[J]. Electromechanical Engineering, 2011, **27**(3): 24-27. (in Chinese)

IOP Conf. Series: Materials Science and Engineering **638** (2019) 012005 doi:10.1088/1757-899X/638/1/012005

[9] Hou Xinbin, Wang Li, Zhang Xinghua, et al. Concept design of multi-rotary joints SPS[J]. Journal of Astronautics, 2015, **36**(11): 1332-1338. (in Chinese)
[10] Liu Zili, Jia Haipeng, Wang Li, et al. Key technologies and inspire on novel space-applied roll ring[J]. Spacecraft Environmental Engineering, 2016, **33**(1): 72-76. (in Chinese)
[11] Yan Minggao, Liu Bocao, Wu Shiping, et al. China aviation materials handbook [M]. 2002: pp470-471. (in Chinese)
[12] Wei P S, Wu T H. Electrical contact resistance effect on resistance spot welding[J]. International Journal of Heat & Mass Transfer, 2012, **55**(s11–12): 3316–3324.
[13] Holm R. Electric contacts[M]. Stockholm, Sweden: Hugo Gebers Forlag, 1964: pp15-31.

Acknowledgement

This work was supported by the State Administration of Science, Technology and Industry for National Defense (Project No: [2017]772).

A novel anti-lock braking strategy for high speed light weight dual-motor electric drive tracked vehicles

Zhaomeng Chen, Xiaojun Zhou and Yuanhe Liu

State Key Laboratory of Fluid Power and Mechatronic Systems, Zhejiang University, 310027, Hangzhou, China

chernzm@zju.edu.cn

Abstract. Anti-lock braking strategies are of great importance in improving manoeuvrability of vehicles. However, studies on that of tracked vehicles are rare compared with wheeled vehicles due to their complicated drivetrain and the sluggish of their braking system. Dual-motor electric drive tracked vehicles (DDTVs), whose drivetrain differs significantly from the traditional ones, have drawn a lot of attention in the trends of hybridization and electrification for tracked vehicles, and made it practical to apply advanced braking control strategies with the application of permanent magnet synchronous motors (PMSMs). This paper develops a novel anti-lock braking strategy with a sliding mode controller and a rule based braking torque allocating method to enhance the braking stability. Simulations are conducted under a typical low adhesion condition to compare the control performances of the proposed strategy with normal full braking strategy and traditional anti-lock braking strategy transplanted from wheeled vehicles. Simulation results show that the proposed control strategy performs better than the other two strategies in reducing stopping distances while still has the ability of regenerate braking energy.

1. Introduction

DDTVs have drawn lots of attentions in the trends of hybridization and electrification of tracked vehicles [1]. The tracks of a DDTV are driven separately by two PMSMs attached to the drive wheels through each semi axle with a mechanical brake installed on. The schematic diagram of it is shown in figure 1.

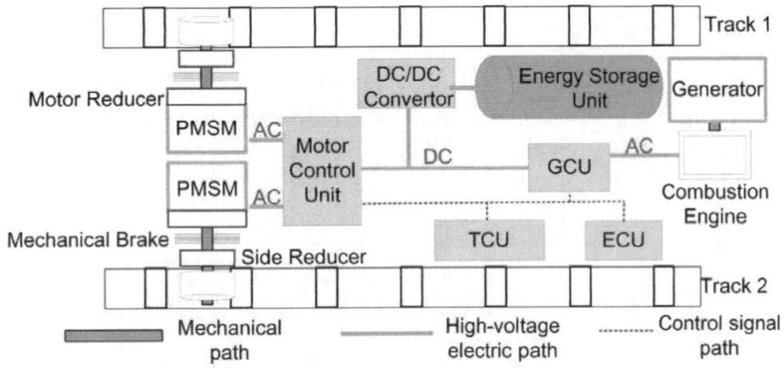

Figure 1. Schematic diagram of a DDTV

IOP Conf. Series: Materials Science and Engineering **638** (2019) 012006 doi:10.1088/1757-899X/638/1/012006

Combined braking system of tracked vehicles normally refers to hydraulic-mechanical braking systems with hydraulic retarders [2]. However, as for high speed light weight DDTVs, PMSMs can replace the role of them and enhance the dynamic response when braking. A strategy called full braking, which means the braking system provides as much braking torque as possible is widely used in tracked vehicles for emergency braking. Even in latest researches of DDTVs, the focus are still on how to coordinate the mechanical brakes and PMSMs produce enough braking torque [3, 4]. However, adopting this strategy will cause the locking of drive wheels and tracks, resulting in a poor braking performance.

To overcome similar drawbacks in the field of wheeled vehicles, traditional ABS system relies on high frequency switch of solenoid valves in braking circuits as well as advanced slip ratio control strategies are studied in previous researches [5-8]. Although research on traction control of tracked vehicles using sliding mode method has been conducted long before [9], to brake stably on a low adhesion surface is still difficult for a tradition tracked vehicle due to the sluggish response of a typical mechanical brake. However, with the participant of PMSMs with fast torque response ability, applying an advanced anti-lock braking control strategy is practical for DDTVs.

The organization of this paper is as follows. In section 2, models related to the braking process of DDTVs are introduced. In section 3, the design of proposed anti-lock braking strategy is introduced. Matlab/Simulink simulations of the proposed control strategy as well as full braking and traditional ABS strategy are shown and compared in section 4. Finally, conclusions are given in section 5.

2. Modelling

The walking structures of a tracked vehicle is significantly different from that of a wheeled vehicle, thus models related to the braking process of a tracked vehicle need to be built specially. Since the features we focused on of PMSMs and mechanical brakes in this case are their torque output capacity and response ability, simple models of first-order process based on experiments is suitable. As a result, what we need to focus on is the modelling of vehicle dynamics, track-ground conditions and driver operations.

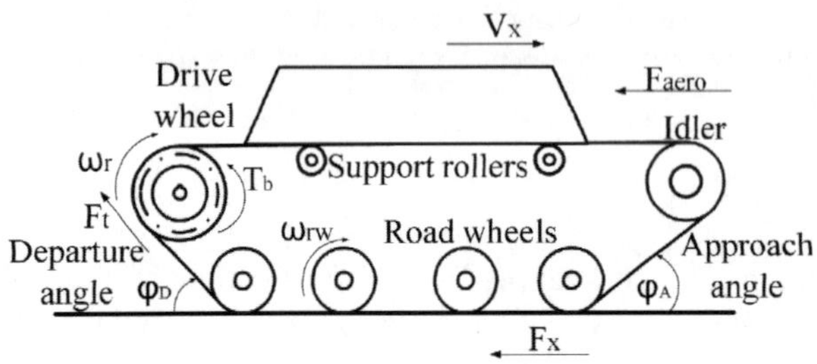

Figure 2. Schematic diagram of forces and torques acting on a DDTV when braking

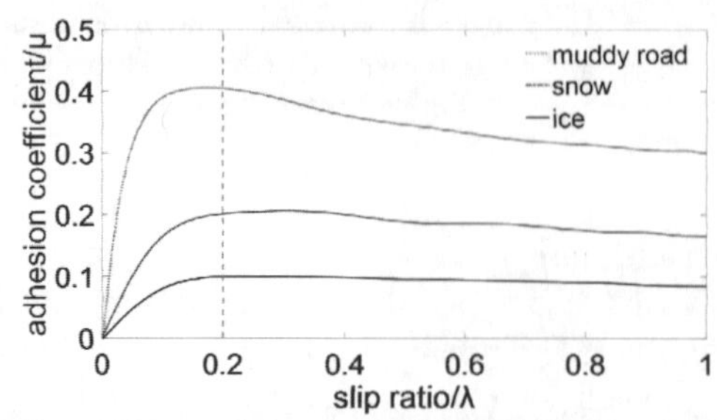

Figure 3. $\mu - \lambda$ curves of three medium and low adhesion track-ground conditions.

2.1. Longitudinal dynamics of a DDTV

As shown in figure 2, the movement of a tracked vehicle can be divided into the rolling of drive wheels, the longitudinal motion of vehicle body, the rotational motion of the tracks, as well as the rolling of road wheels, idler and support rollers. Assuming that the track is uniform and soft, and there is no slip or slide between the track and any wheel, a semi-vehicle longitudinal dynamic model in braking situation can be established. Similarly with wheeled vehicle, the longitudinal motion of vehicle body and the rolling of drive wheel can be denoted as:

$$\begin{cases} F_x + F_{aero} + M\dot{v}_x = 0 \\ J\dot{\omega} - F_t r + T_b = 0 \end{cases} \tag{1}$$

where F_x is the track-ground resultant force of one track, F_{aero} is half of the air drag, M is half of the DDTV's total mass, v_x is the longitudinal velocity of the vehicle, J is the rotational inertia (RI) of drive wheel and transmission system attached to it, ω is the rotational speed of the drive wheel, F_t is the force applied on drive wheel by the track, r is the radius of the drive wheel, and T_b is the braking torque applied on drive wheel by the transmission system.

In order to describe the motion of the track and other wheels attached to it, a method of calculating the equivalent rotational inertia (ERI) based on energy conservation theory is proved effective in [10, 11]. Taken the slip ratio defined in equation (2) into consideration, the ERI of track and wheels when braking can be expressed as equation (3).

$$\lambda = (v_x - \omega r)/v_x \tag{2}$$

$$\begin{cases} J_{eri}^{wi} = n_{wi} J_{wi} r^2 / r_{wi}^2 \\ J_{eri}^{t\_f} = m_{t\_f} r^2 [1 + 1/(1-\lambda)^2 - 2\cos\varphi_A/((1-\lambda)] \\ J_{eri}^{t\_r} = m_{t\_r} r^2 [1 + 1/(1-\lambda)^2 - 2\cos\varphi_D/(1-\lambda)] \\ J_{eri}^{t\_u} = m_{t\_u} r^2 [(2-\lambda)/(1-\lambda)]^2 \\ J_{eri}^{t\_b} = m_{t\_d} r^2 [\lambda/(1-\lambda)]^2 \end{cases} \tag{3}$$

where $J_{\omega i}$ is the RI of the i th kind of wheel, $n_{\omega i}$ is the quantity of this kind of wheel, and m_{t_f}, m_{t_r}, m_{t_u} , m_{t_b} are the masses of the front part, the rare part, the upper part as well as the bottom part of the track respectively. Thus, the longitudinal dynamics of the DDTV can be amended by replace J with J_{eri} which is the summation of equation (3) and J in equation (1).

2.2. Model of track-ground

Due to the differences between the contact form of track-ground and tire-road, there are no suitable analytical models of track-ground for tracked vehicle. Thus, the best way is establishing an experimental model. Some medium and low adhesion track-ground conditions are shown in figure 3.

According to figure 3, we can find that when the slip ratio nears 0.2, the adhesion coefficient achieves the largest value, thus the value 0.2 can be called the optimal slip ratio.

2.3. Model of driver operation

The travel of brake pedal can be interpreted as a demand deceleration and it is usually considered as meaningful after a small idle motion, as well as deemed to need a full braking when nearing the terminal. Thus the demanded deceleration can be defined proportional to the effective pedal travel.

3. Design of anti-lock braking strategy

According to the national military standard of China, the maximum deceleration must be higher than $5 m/s^2$ and it is usually set as $5.5\ m/s^2$ [4]. As shown in figure 3, it's impossible to reach such deceleration when the vehicle runs on a low adhesion ground and will leads to the locking of tracks and wheels if braking torque remains unchanged. Thus, a sliding mode strategy which calculates demand braking torque in real time can overcome this problem.

Substituting λ and $\dot{\omega}$ into the derivation of λ we can get that:

$$\dot{\lambda} = [(1-\lambda)^2 J_{equ} \dot{v}_x + (1-\lambda) F_x r^2]/[\omega r J_{eri} + (1-\lambda) T_b / \omega J_{eri} \tag{4}$$

A zero-order sliding surface can be chosen to track the demanded slip ratio and the sliding surface and its derivation with respect to time is:

$$\begin{cases} s = \lambda - \lambda_d \\ \dot{s} = \dot{\lambda} - \dot{\lambda}_d \end{cases} \tag{5}$$

In order to converge quickly when s is large and reduce chattering when s is small, a sliding mode control method using exponential approaching law is suitable for this control system [12]. Considering that λ_d is the optimal slip ratio which is a constant, the derivation of sliding surface can be written as:

$$\dot{s} =- \varepsilon sign(s) - ks = \dot{\lambda}, \quad \varepsilon > 0, \quad k > 0 \tag{6}$$

where sign(s) is symbolic function of s, k should be designed relatively large and ε should be relatively small. Thus, the total demanded braking torque converted to drive wheel can be derived as:

$$T_{bd} = rF_x + (1-\lambda) J_{eri} \dot{v}_x / r + \omega J_{eri} [\varepsilon sign(s) + ks]/(1-\lambda) \tag{7}$$

However, the total demanded braking torque changes rapidly in the braking process, the performance will meet a significant degradation if there doesn't has a proper allocating method. The braking system of a light weight DDTV consists of mechanical brakes and PMSMs and these actuators have their own advantages and drawbacks. The mechanical brake can offer a large braking torque, but its dynamic response is quite poor. As for PMSM, its torque response speed is much faster than mechanical brake, however, the maximum braking torque it can provide declines with the increasing of revolution speed when rotating faster than its rated speed, even if its rated torque is already much smaller than mechanical brakes. What's more, the performances of the PMSM also get worse when the revolution speed is low.

Therefore, several fundamental allocating rules could be set as: 1.the total braking torque should be divided into a steady part and a dynamic part; 2.the steady part is provided by mechanical brake and the dynamic part is provided by the PMSM; 3.the PMSM exits braking process when the vehicle speed is slower than 5 km/h; 4. the PMSM does not output any positive torque when braking; 5. to supplement the shortage of the dynamic part of total demanded braking torque when s is large, a compensating coefficient is introduced to compensate the difference of them using mechanical brake.

Thus, substituting the definition of sliding surface into equation (7) can get that

$$\begin{cases} T_{bds} = rF_x + (1-\lambda_d)J_{eri}\dot{v}_x/r + \chi(T_{bdd} - T_{mo}) \\ T_{bdd} = \omega J_{eri}(\varepsilon + F_d)sign(s)/(1-\lambda) + J_{eri}(-\dot{v}_x/r + \omega k/1-\lambda)s \end{cases} \tag{8}$$

where T_{bds} is the steady part of total demanded braking torque, while T_{bdd} is the dynamic part. T_{mo} is the real braking torque of PMSM converted to the drive wheel and χ is the compensating coefficient. To reduce chattering, the symbolic function can be replaced by a saturation function $sat(s/\Phi)$.

4. Simulations

Simulations are taken with the proposed control strategy as well as two other strategies to compare the braking performance and verify the effectiveness. All the three strategies are separately:

•**FB**: which is full braking strategy. Mechanical brake and PMSM are working simultaneously to provide enough braking torque to achieve demand deceleration despite the track-ground condition;

•**TABSC**: which transplant a traditional ABS with PWM driven solenoid valves added in wheel cylinders. When it is triggered, PMSM exits and all the braking torque is provided by mechanical brake.

•**SMSC:** which is the proposed control strategy that mechanical brake and PMSM work simultaneously.

Table 1. Main parameters of the DDTV for simulation

Parameter	Value	Parameter	Value
Vehicle mass, m(t)	32	Approach angle, φ_A(°)	27.3
Drive wheel radius, r(m)	0.309	Departure angle, φ_D(°)	35.6
Frontal area of the vehicle, A(m^2)	5.36	Ratio of the side reducer, ir	4.59
Aerodynamic drag, C_D	1	Ratio of the motor reducer, im	2.2
Air density, ρ(Ns^2/m^4)	1.22	Rated power of the PMSM, P_m(kW)	375
RI of drive wheel & powertrain,(kgm^2)	158.8	Rated speed of the PMSM, n_m(rpm)	3000
RI of idler,(kgm^2)	31.2	Max speed of the PMSM, n_{max}(rpm)	9000
RI of road wheel,(kgm^2)	23.7	Quantity of road wheels, one side	6
RI of support roller,(kgm^2)	13	Quantity of support rollers, one side	2

Five indicators are chosen to evaluate the performances which are stopping time t, stopping distance S, mean value of slip ratio $\bar{\lambda}$, standard deviation of slip ratio σ and regenerated energy E.

The simulations are taken in the environment of Matlab/Simulink R2015b and the schematic diagram of the model is shown in figure 4. The DDTV is braked at a speed of 75km/h on snowfield and the brake pedal is set to increase from 0 to 100% in 10 ms half a second after the accelerator pedal decreased to 0. The start point of plotting and calculating the indicators is defined at when accelerator pedal started to decrease. The main parameters of the DDTV for the simulations are listed in table 1.

Simulation result with the proposed strategy is shown in figure 5, and the effect of these strategies can be reflected in table 2. In figure 5, VV, TV, MBT, MRT and TBT are the abbreviations of vehicle velocity, track velocity, mechanical brake torque, motor regenerative torque and total braking torque respectively.

Table 2. Value of the indicators

	t(s)	S(m)	$\bar{\lambda}$	σ	E(kJ)
FB	9.78	101.86	0.8413	0.3050	516.29
TABSC	9.01	97.24	0.1674	0.1087	0
SMSC	8.65	92.19	0.1894	0.0396	316.7

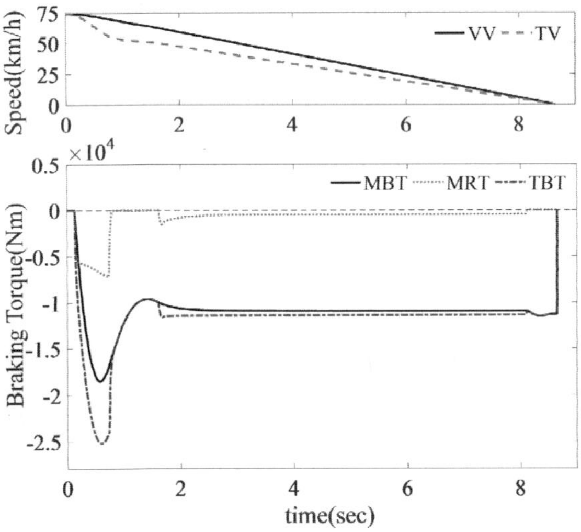

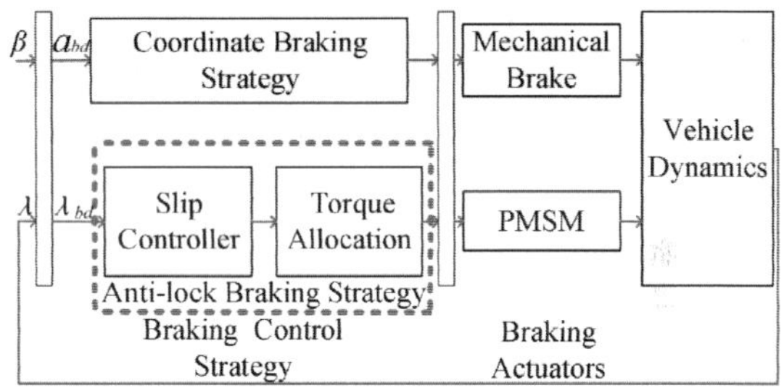

Figure 4. Schematic diagram of simulation model

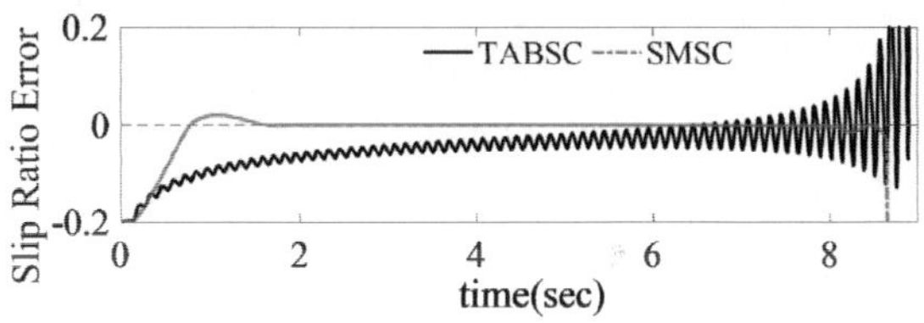

Figure 5. Simulation results of emergency braking on snow

Figure 6. Error of slip ratio using TABSC and SMSC

It is obvious that whether transplanting a traditional ABS or applying the proposed control strategy can improve the braking performance, however, applying the proposed strategy end up with a much larger improvement. The reason can be found in figure 6 which shows the error of real slip ratio and the optimal one using TABSC and SMSC respectively. We can easily find the slip ratio of SMSC converges faster than that of TABSC without such oscillation. Hence, applying SMSC method can make the most use of the ground adhesion to decelerate the DDTV as the result shown in table 2.

5. Conclusion

The proposed anti-lock braking strategy utilizes the advantages of mechanical brake and PMSM, enables the braking system providing braking torque rapidly and accurately. When compared with FB, the stopping distance reduces by 9.49%. When compared with TABSC, the proposed one still has an advantage of reducing it by 5.19%. Although using the proposed strategy recovers less energy than using the FB strategy, it is negligible compared with the significant advances in other aspects. What's more, applying the proposed strategy doesn't have to change the original braking circuit of the DDTV, which can save the budget of adding solenoid valves of each wheel cylinder. In conclusion, applying the proposed anti-lock braking strategy can achieve best result with the minimal cost.

IOP Conf. Series: Materials Science and Engineering **638** (2019) 012006 doi:10.1088/1757-899X/638/1/012006

References

[1] Yuan Z, Fengchun S, Chengning Z 2007 *J. B. Inst. Techno.* **27** 29-33
[2] Guangjun Z, Jiangang L 2010 *CMCE (Changchun)* pp 515-518
[3] Hui S, Jiangtao G, et al 2017 *Acta. Armamentarii.* **38(5)** 1027-1034
[4] Hui S, Jiangtao G, et al 2017 *J. TYUT* **48(1)** 79-85
[5] Patra N, Datta K 2012 *ICACCCT(Ramanathapuram)* pp 385-391
[6] Hamzah N, Yahay M, Selamat H, et al 2012 *CSPA(Melak)* pp 138-143
[7] Jonathan M, David C 2010 *Vehicle. Syst. Dyn.* **48** 373-392,
[8] Bera K, Bhattacharya K, Samantaray K 2011 *P. I. Mech. Eng. J-J. Eng.* **225** 918-934
[9] Zhejun F, Yoram K, David W 1995 *ACC(Seattle)* pp 1176-1177
[10] Jiajun Y, Xiaojun Z 2013 *J. Vib. Shock.* **32** 68-72
[11] Zhe W, Haoliang L, Xiaojun Z, Zhaomeng C, Yong Y 2018 *Sensors.* **18(7)**, 1993-.
[12] Jinkun L 2017 *Sliding mode control using MATLAB.* (Amsterdam: Elsevier)

A Position Synchronization Control for Dual Redundant Electro-mechanical Actuation in Flight Control System

D Y Wang[1], D Q Li[1], J Chen[2] and S C Zheng[2]

[1] Department National Key Laboratory of Science and Technology on Aerospace Intelligence Control, BJ 100854, China
[2] Beihang University, BJ 100091, China

deyiwang@vip.qq.com

Abstract. The paper develops a position synchronization controller for a new active/active dual redundant electro-mechanical actuation system and designs a test platform to validate the performance of the controller. The mathematic models of the position synchronization system are firstly built based on the schematic of the active/active dual redundant electro-mechanical actuation system. The position controller for DEMA and GEMA of is developed based on the velocity and acceleration feed forward and traditional PID theory. A series of motion experiments are carried out to verify the effectiveness of the controller design for active/active dual redundant electro-mechanical actuation system. The experiments results show that the position controllers designed for either DEMA or GEMA are satisfied, the biggest following error is not more than 0.1mm. And the synchronous motion tests show that the PI plus Output Saturation feedback position synchronization control can reduce more than 30 percent synchronous position errors. Furthermore, the tests of position synchronization show that EMA stiffness has a big influence on the synchronous precision. The new position synchronization controller designed based on the accurate system model and actual system parameters can better achieve the position synchronization requirements of the dual redundant EMA system, which makes good sense for redundant systems.

1. Introduction

One hundred years ago, on December 17, 1903, the Wright brothers, Orville and Wilbur, invented and built the world's first successful aircraft and made the first controlled, powered and sustained heavier-than-air human flight [1]. From then on, human have produced thousands of various aircrafts, which have greatly changed the world. The aircraft is a complex vehicle flying on air and composed by three main parts: airframe, engine and on board systems [2]. The flight control system belongs to this on board systems and is powered by the hydraulic and electric ones. It is divided into primary level and secondary level. It is in charge of controlling the aircraft attitude (the yaw axis by rudders, pitch axis by elevators and roll axis by ailerons) and aerodynamic configuration (the high lift by slats and flaps, trim by trimmable horizontal stabiliser and the air brake spoilers and some special surfaces) by moving the flight control surfaces [3][4].

Due to the maturity and power density of the hydraulic technology, conventional hydraulic power has been widely used to actuate the actuators in the past 50 years. In this technology, the power energy is given to the fluid by increasing its working pressure (now up to 35MPa), and is transmitted by mass transfer (Power-by-Pipes, PbP). This has a good advantage of taking away the heat generated by power losses, however, this induces many engineering constraints and makes the maintenance cost keep high. The current development aim in aviation is to make large commercial aircraft be cheaper,

IOP Conf. Series: Materials Science and Engineering **638** (2019) 012007 doi:10.1088/1757-899X/638/1/012007

safer and greener. The more electrical aircraft (MEA) can offer mangy potential advantage in terms of performance, maintenance, integration, reconfiguration, ease of operation and management of power[5]. Electrical powered (Power-by-Wire, PbW) actuation system is considered by aircraft makers as an important means to achieve the MEA goals. Electro-mechanical actuation (EMA) as one of the most important technology of electrical powered actuation (EPA) system is introduced in the new generation of aircraft in replacement of the conventional PbP actuation [6] and is gradually used widely. In 2011, Boeing not only introduced electro-mechanical actuator (EMA) for spoilers (4 out of 14) for the B787 to fulfil the spoiler actuation function in case of loss of hydraulic power, but also for the THS [7]. Also in 2011, another big aircraft manufacture Airbus introduced EMA for a civil aircraft aileron on the A320 MSN1 F2675. Until June, 2012, the EMA has successfully completed 38 flight cycles and 114 hours without any failure alarm [8]. As electro-hydraulic actuation (EHA) is considered as the transitional form of EPA to EMA, the EMA will be introduced more and more into the primary flight control system.

As a safety critical application, the primary flight control system is strictly required to have a very good performance on statics, dynamics and reliability. For the primary flight control system, redundant actuators are generally employed to drive a common surface; for the secondary flight control system, one actuator with internally redundant arrangements is used for each function [9]. The existing redundant actuation configurations and their properties and applications are Parallel Force/Position Summing, Tandem Force/Position Summing and Speed Summing.

The parallel force/position summing configuration is a common choice for the primary flight control system on commercial airliners. But in most cases they operate on active/active mode, the force fighting between actuators in parallel force summing configuration is serious and must be addressed. However, there is no force fighting in the parallel position summing configuration [9] which is suitable for all kinds of actuator. In this research, the redundant actuators composed by two different EMAs are selected to be studied in present research.

This thesis is proposed to address the position synchronization issue of a hybrid electro-mechanical actuation system. The test platform of the system is designed. In section 2, the motion of the redundant system is analysed, and then the nonlinear mathematic model of the system is built. The schematic of the synchronization strategy is given. In section 3, the virtual model of the system is built in the AMESim virtual environment, and the performance of the synchronization strategies is pre-validated in this virtual environment. In Section 4, the experimental design and parameter setup is given. In section 5, a series of motion experiments are carried out. The performance of the controller with and without synchronization strategies are verified by the synchronous motion experiments. The influences of system stiffness of the redundant system are also investigated. The present work is concluded and an outlook over the further improvements and perspectives is provided in Section 6.

2. Modelling and analysis of the synchronous motion system

In order to model the position synchronous configuration, first, models for the two actuators are presented, which will be connected with a model for the control surface structure. The schematic of synchronous motion system is presented in Fig. 1.

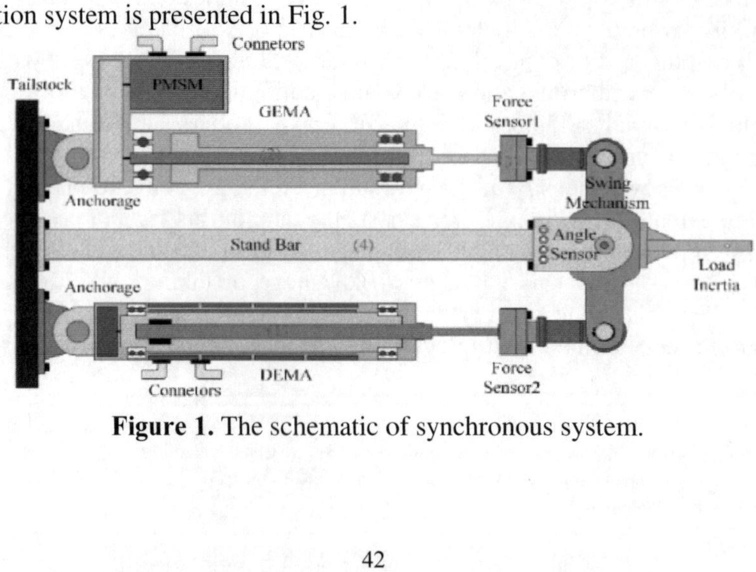

Figure 1. The schematic of synchronous system.

The synchronous motion system researched in this paper includes two EMAs which combine PMSM motor driving an inverted planetary roller-screw (PRS). The system is balanced with both EMAs working together as one EMA is driven to extend the other is counter driven to retract. The system is generally in one of three conditions: ACTIVE, when the system is in operation and STAND-BY when the control system is powered-up but the motors are isolated or AUXILIARY when only one EMA is actuating the flight control surface. The two employee are different types of EMA. Depending on the type of drive, linear EMAs can be divided into two types: gear drive and direct drive, shown in the Fig. 2.

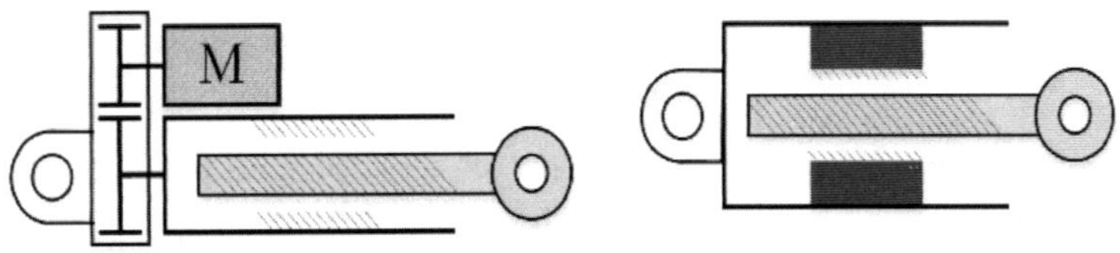

Figure 2. Type of the linear EMA.

The gear drive type of linear EMA comprises a single PMSM motor that drives through a transfer spur gear train, (1:1 ratio gearbox that is incorporated due to the size envelope restrictions). The gearbox is lubricated and sealed for the duration of the proposed trial fit programme. The gear train drives into a planetary roller screw and by earthing the roller screw nut assembly the rotational energy is transferred into linear force acting on the output rod. This structure needs all the separate parts to be under good performance. This can be used in the field without space limit. However, the direct drive EMA eliminates the gear reducer and timing belt. The screw nut is compacted with motor rotor. This design makes the EMA be much smaller and more independent than traditional rotary to linear conversion mechanisms. The linear EMA in the Fig. 3 is an actual product made of Exlar company. Other company like SKF can also produce this kind of EMA.

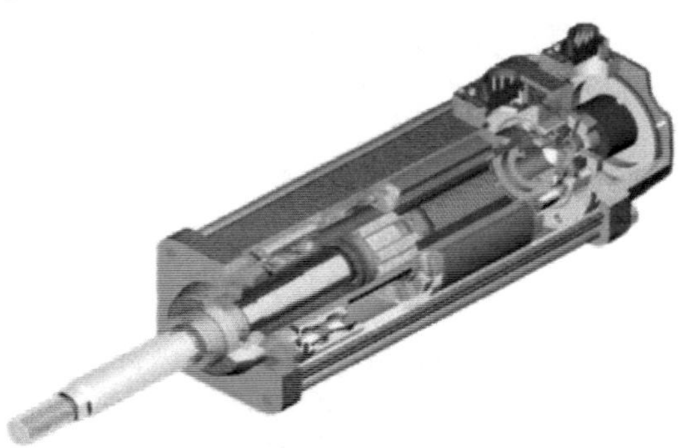

Figure 3. Linear actuators with integrated motor.

The reversible planetary roller screw (PRS) is preferred to ball screw in the compact EMA, firstly because the roller screw has higher loading carrying capacities and better stiffness. For another reason, the PRS has a near 15 times greater life time than ball screw, with the same other condition. Though the PRS EMA has many advantages, its structure becomes complicated. There are many work still need to be done, such as system modelling, friction analysis.

2.1. Modeling of the electro-mechanical actuation

As is shown, in the Fig. 4, the model of EMA comprises a nonlinear description of a velocity controlled PMSM, while inner loop is a current controlled loop, and a model of mechanical transmission kinematics including a gear box (or direct drive) and a roller-screw [10-11] actuator, with position and speed controlled in the outer loop.

IOP Conf. Series: Materials Science and Engineering **638** (2019) 012007 doi:10.1088/1757-899X/638/1/012007

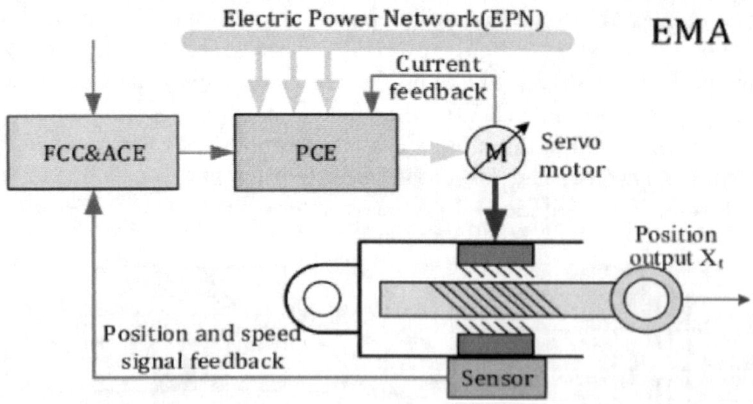

Figure 4. Schematic of EMA.

For the description and vector control of the motor dynamics, current and voltage are transformed from the 3-phase system to the direct and quadratic axis of the rotor by the Park-transformation. The transformation matrix is written as [12-13]:

$$C_{\frac{3s}{2r}}(\theta) = \begin{bmatrix} \cos\theta & -\sin\theta & 1 \\ \cos(\theta-120^0) & -\sin\left(\theta-120^0\right) & 1 \\ \cos(\theta+120^0) & -\sin\left(\theta+120^0\right) & 1 \end{bmatrix} \tag{1}$$

Electromagnetic torque equation of the BLDM can be written as:

$$T_e = \frac{3}{2} n_p \Psi_r i_q + \frac{3}{2} n_p \left(L_d - L_q\right) i_d i_q \tag{2}$$

Where L_q, L_d is the inductance in q-axis or d-axis, i_d, i_q is the current in q-axis or d-axis, n_p is the number of pairs of poles t, and Ψ_r is the magnetic flux of the permanent magnet. Considering the nonlinear friction and external disturbances, the torque balance at motor shaft can be derived as:

$$T_e + T_d(t) = T_L + J\ddot{\theta}_m + T_{fm}(\omega_m) \tag{3}$$

Where T_e is the motor electrical output torque $[N \cdot m]$, T_d is total external disturbance $[N \cdot m]$, T_L is the effective torque on the screw $[N \cdot m]$, T_{fm} is the nonlinear friction torque $[N \cdot m]$ including static friction torque, Coulomb friction torque and Stribeck effect friction torque, J is total rotary inertia in motor side $[kg \cdot m^2]$, θ_m is motor angular position [rad] and ω_m is motor angular velocity [rad/s].
As is known, roller-screw transfers the rotational power into translational form though:

$$k_{ema} = 2\pi / P_{rs} \tag{4}$$

Where k_{ema} is drive ratio and P_{rs} is roller-crew lead [mm].
Considering the direction dependent efficiency η_{ema}, which represents the efficiency of the gearbox and the roller-screw, and external load force F_{ema} [N], the torque balance at roller-screw can be derived as:

$$F_n = m_{ema}\ddot{x}_{ema} + F_{ema} \tag{5}$$

Where m_{ema} is equivalent mass of the roller-screw and surface [Kg] and F_n is the roller-screw input force, which can be written as:

$$F_n = \eta_{ema} k_{ema} T_L R \tag{6}$$

Where R is gearbox ratio [dimensionless].

The Laplace transformation of upper seven equations is done to get the prototype of EMA, as shown in Fig. 5.

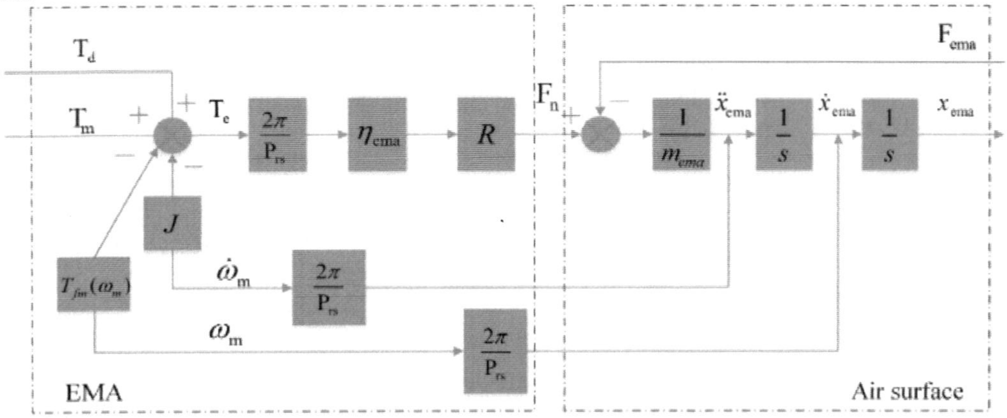

Figure 5. Block diagram of EMA.

In Fig. 5, it can be found out that the EMA system is structured by two parts. One part is the EMA which is made up of one PMSM and one reducing mechanism roller-screw. The inputs of this part are the reference torque command T_m , the rod moving speed $\dot{x}_{ema}$ and acceleration $\ddot{x}_{ema}$. And the output is the force, F_n generated by EMA. Another part is the air surface which can influence the aerodynamic characteristics of airplane, the inputs of this part are the EMA output force F_n and the external load force F_{ema}, and the outputs are the air surface moving displacement x_{ema}, speed $\dot{x}_{ema}$ and acceleration $\ddot{x}_{ema}$. The joint between rod and the air surface can be looked as rigid, so the displacement, speed and acceleration of rod are equal to the ones of air surface.

2.2. Modeling of control surface structure

The control surface structure is modelled as a mass, representing the surface inertia , and equal attachment stiffness and damping for two EMAs. The inputs are EMAs displacements and , as well as the external air loads . The outputs are the control surface angel and the forces acting on the EMAs and . The equation is thus derived as:

$$J_{cs}\ddot{\theta}_{cs} = F_{gema}R_{ema}\cos\theta_{cs} + F_{dema}R_{ema}\cos\theta_{cs} - F_{air}R_{cs} \quad (7)$$

$$F_{gema} = c_{cs}(x_{gema} - x_{cs}) + d_{cs}(\dot{x}_{gema} - \dot{x}_{cs}) \quad (8)$$

$$F_{gema} = c_{cs}(x_{gema} - x_{cs}) + d_{cs}(\dot{x}_{gema} - \dot{x}_{cs}) \quad (9)$$

Where $x_{dema} = x_{gema} = R_{ema}\sin\theta_{cs}$, $x_{cs} = R_{ema}\cos\theta_{cs}$.

The equation(7), (8) and (9) show that surface stiffness has an influence on the kinetic characteristic. In fact, the system stiffness bring lots of problems for the redundant system, like force fighting, out of synchronization and limit cycle. The experiments carried out in the next section shows something on this aspect.

2.3. Modeling of synchronous motion system

The controller of the synchronous motion system is realized by IPC, which can be used to simulate the FCC (flight control computer). The outputs range of the force and encoder sensors are +/-10VDC. They are collected by Target Controller. The output rang of IPC board are also +/-10VDC, which are calculated by the IPC basing on the synchronous control strategy. The position command is separately sent to GEMA and DMEA controllers. The GEMA and DEMA controller complete the GEMA and DEMA control. The analog output from Target controller includes velocity, position and torque demand. The current and velocity message are sent back to the Target controller, as well as errors and status information. The schematic block of the synchronous motion system is shown in the Fig. 6.

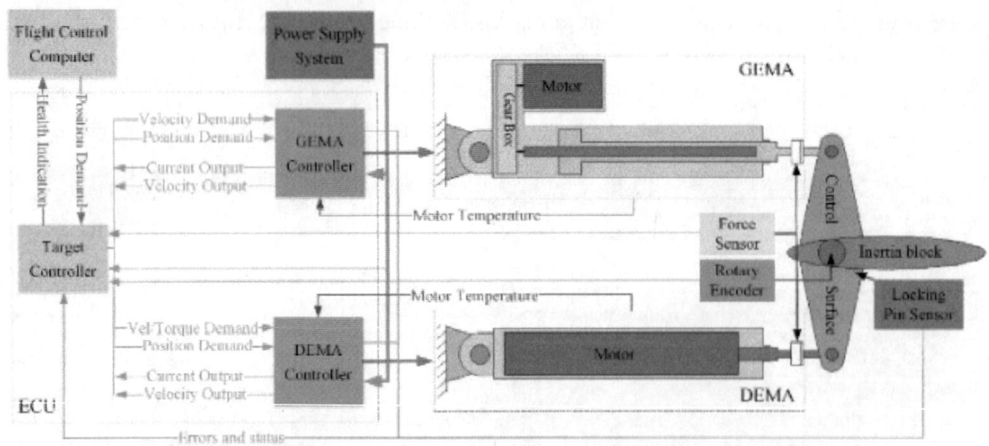

Figure 6. The schematic block of the synchronous motion system.

Based on the models of the single channel of EMA and control surface, the dynamic mathematic model of the synchronous motion system of EMAs can be derived. The structure diagram of the synchronous motion system without controller is shown as Fig. 7.

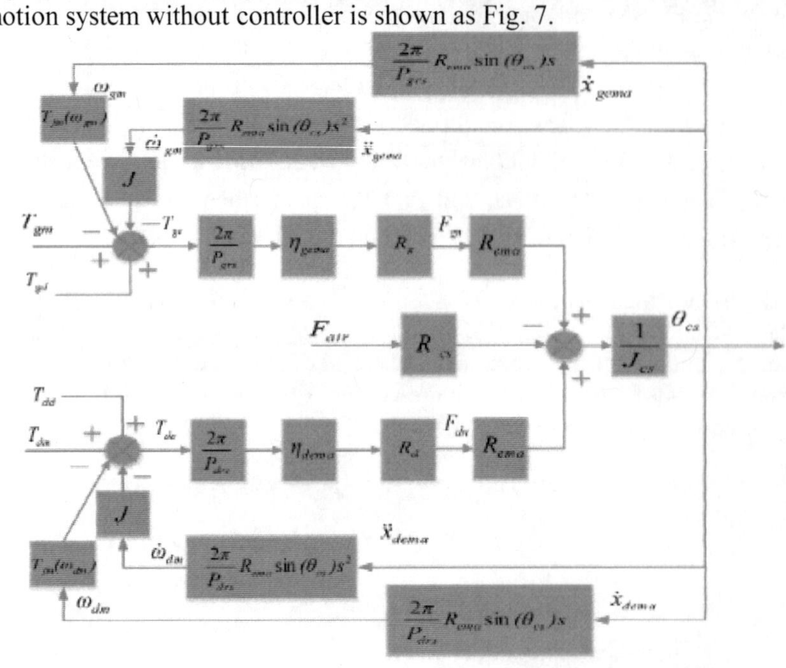

Figure 7. Structure diagram of synchronous motion system.

It is shown in the Fig. 7 that there are three parts in this synchronous motion system, EMAS part, Compliance part and control surface part. Based on the different parameter, every EMA has its own controller, the controllers are designed and validated in the following tests. In the compliance part, the compliances between EMA and control surface are assumed as the same, but because of the differences of the structure and mechanical connection of the EMA, the value is different. And this has a big influence on the system synchronization accuracy, which will also be validated in the following test. External variables are target position command X_r and external load F_{air}.

3. Experiment setup

For carrying out the research, a hybrid actuation test bench composed one DEMA and one GEMA is designed and employed. The test bench was built for evaluating the motion behaviors of the two EMAs and placed in the Fluid Power Transmission and Control Research Center of SMEA, BUAA

Beijing, China, as shown in Fig. 8. The test bench will be used to demonstrate the performance of redundant actuation system in laboratory environment as the level 4 of technology readiness level. The position synchronization control force strategies to be proposed will be validated on this test bench.

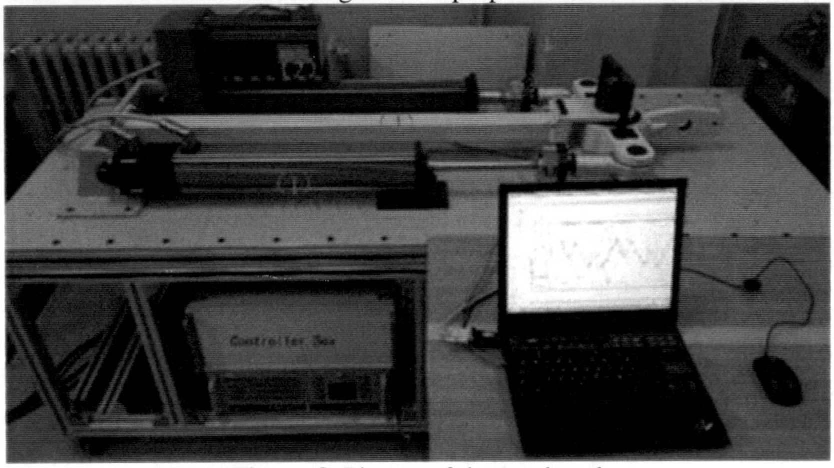

Figure 8. Picture of the test bench.

3.1. Mechanical Configuration

The test bench includes one DEMA and one GEMA which both combine a PMSM motor driving a planetary roller-screw (PRS). There is a gear box in the GEMA driving screw to do rotary motion, while the DEMA eliminates the gear reducer, the screw nut is compacted with motor rotor, as is shown in Fig. 8 and Fig. 1. One swing mechanism with two joint bearing connecting end is used to connect the two EMAs. This swing iron lump can rotates around the central axis, driven by these two EMAs. The ellipse iron lump is used to the surface, especially the inertia. The stand bar between the EMAs is used for supporting the rotating shaft and tailstock. The EMAs are fixed on the bench by two anchorages, which can simulate the linkage compliance to airframe and their value can be adjusted based on requirements. In addition two force sensors are installed on the rod end of actuators to measure their output forces, and angle sensor is stalled on the rotating shaft. The two EMA are fixed in symmetrical position, one push and one pull to position the load, like the parallel arrangement, the system performances are quite similar. The results got on this test bench can be applied on other parallel arrangement redundant actuation system. Meanwhile, this arrangement can also be used to study the single actuator's property. In case one actuator is studied, the other one can be force controlled to simulate the air load. The following table summarizes the main properties of test bench:

Table 1. Properties of practical emas.

GEMA		DEMA		Load
Stroke	200mm	Stroke	200mm	Equivalent Mass at the rod: 20Kg
Max Speed	149mm/s	Max Speed	127mm/s	
Max Force	47KN	Max Force	17KN	
Supply Voltage	380V/12A	Supply Voltage	380V/6A	

3.2. Electric Architecture

The schematic of electric system is presented in Fig. 6. The IPC connected to test platform is not inside the control loop. It is in charge of some offline programming, parameter setting and data collection. The control laws are designed in the Matlab/Simulink simulation environment on IPC, and then downloaded to be executed on the target controller. Meanwhile the experiments results stored in target controller target can be uploaded to IPC. All the data transmission between target controller and IPC is on the basis of TCP/IP protocol. And the data transmission between IPC and controller is on the basis of RS232 protocol.

IOP Conf. Series: Materials Science and Engineering **638** (2019) 012007 doi:10.1088/1757-899X/638/1/012007

3.3. System performance Requirements

For the studied active/active position control redundant actuation system, the objective of thesis is designing the position controllers with synchronization for DEMA and GEMA to ensure that the following performances can be obtained:

- Position controller of each EMA performance: position tracking error of each EMA is smaller than 0.5mm;
- Position synchronization controller performance: position synchronization error of the actual position is smaller than 0.5mm;
- Segregation: the cross links between channels should be minimized and limited to ensure the immunity against failure propagation from one channel to another channel;
- Complexity: the strategy needs short computation time and few numbers of sensors.

4. Test result and analysis

4.1. DEMA controller performance

To valid the DEMA controller performance, two groups of experiments have been done. The frequency are 1Hz and 5Hz, and the amplitude utilized decreases from 3mm to 1mm. The schematic of the DEMA controller is shown in the Fig. 9. The velocity and acceleration feed forward, as well as PID are served in the DEMA controller. The proportional, integral and differential gains are 26.87, 296.3 and 0.001, while the velocity and acceleration gains are 0.746 and 0.006.

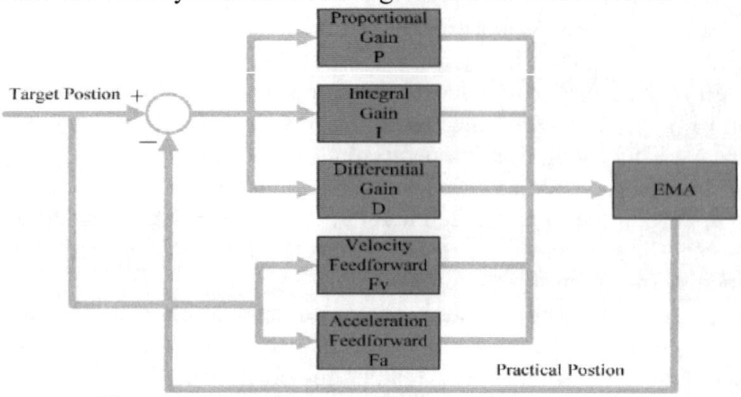

Figure 9. The schematic of the DEMA controller.

The sine position tracking performance of the controller obtained from the test is shown in the Fig. 10. In the Fig. 10 above, the position error following the sine target position (3mm, 1Hz) is nearly zero, the result is very good. However, with the frequency increased to 5 Hz, the following error come about, up to 0.1mm. To be sure, this follow error is acceptable, compared to the system synchronization requirements 0.5mm.

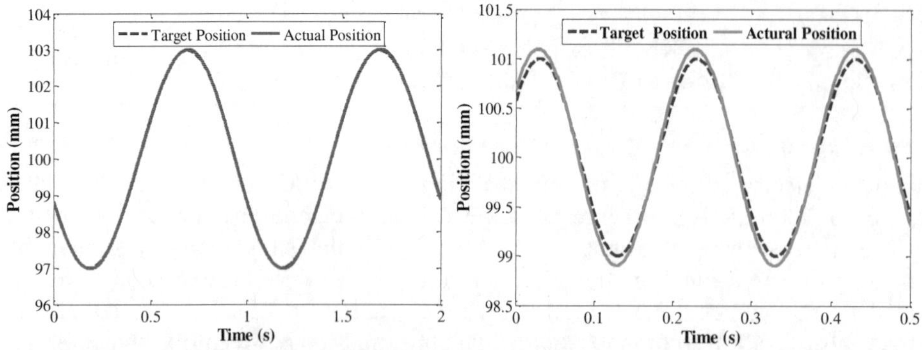

Figure 10. Sine tracking curve of DEMA under different frequency and amplitude.

IOP Conf. Series: Materials Science and Engineering **638** (2019) 012007 doi:10.1088/1757-899X/638/1/012007

4.2. GEMA controller performance

To valid the GEMA controller performance, two groups of experiments have also been carried out. The sine target position signal is the same as DEMA. Because there are many differences between GEMA and DEMA in the aspect of transmission mode, mechanical and electrical parameters, the different controller gains are chosen in FEMA controller. The proportional, integral and differential gains are 22, 202.8 and 0.002, while the velocity and acceleration gains are 0.6687 and 0.0025. Under the different sine signal, 1mm amplitude and 5Hz frequency, 3mm amplitude and 1Hz frequency, the performance of the FEMA is satisfied, shown in the Fig. 11.

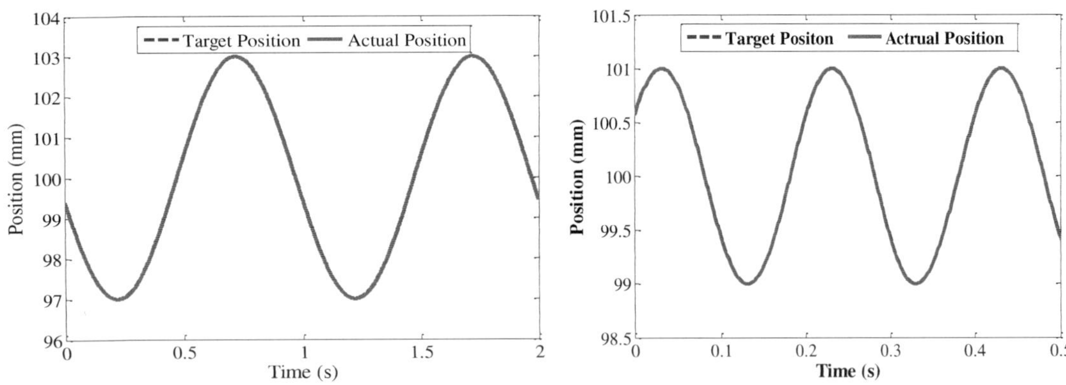

Figure 11. Sine tracking curve of GEMA under different frequency and amplitude.

As is shown in the Fig. 11 (below), the actual following error is small that it can only discovered under enlarged view.

4.3. Position synchronization controller performance

The schematic of the position synchronization motion system without synchronization controller is shown in Fig. 12.

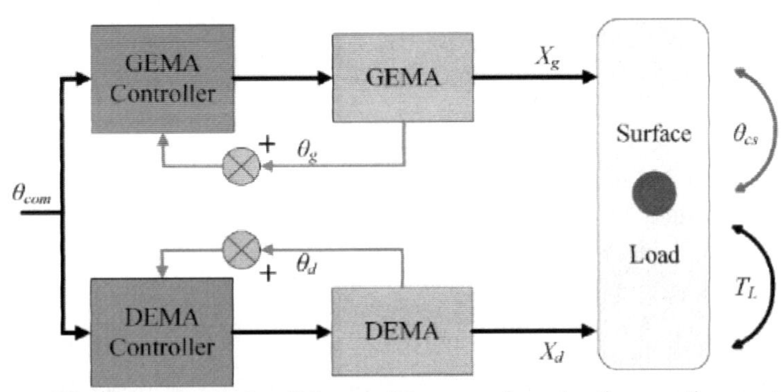

Figure 12. The schematic of the position synchronization motion system.

Depending on the good control performance of the GEMA and DEMA controllers, the actual position tracking result of the two EMAs, without synchronization controller, are shown in Fig. 13. The 10mm position command is separately given to these two EMAs at the same time. From this plot, it shows that DEMA and GEMA almost move together. The asynchronization obviously happens at the beginning of stretch out and draw back moment. The static synchronization errors around 0mm position are unstable, while 10mm position is relatively constant.

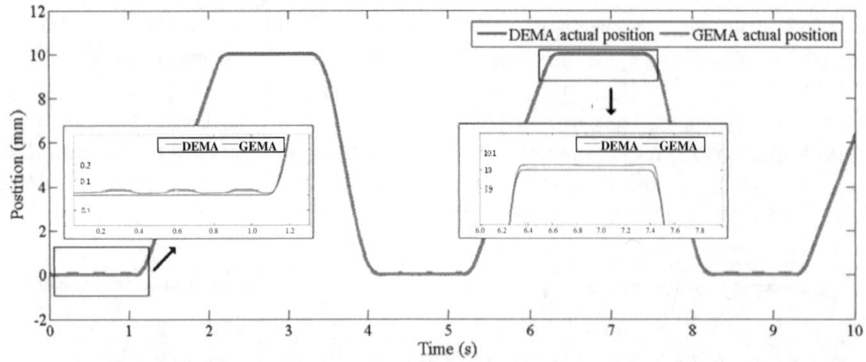

Figure 13. Position synchronization curve without synchronization controller.

The actual position difference between these two EMAs is shown in the Fig. 14. The biggest synchronization error is nearly 0.03 mm, refer to the red curve in Fig.14. From the Fig. 14, we discover that the tracking error from the DEMA (blue curve) make a big contribution to the synchronization error, as much as above 90 percent. However, the tracking error from GEMA is so small that nearly can be ignored.

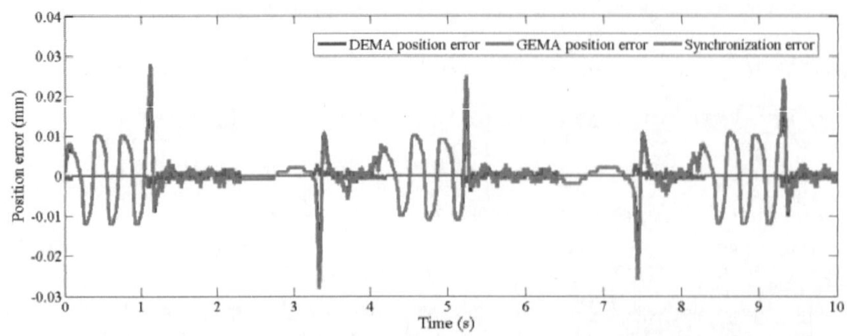

Figure 14. Position synchronization errors without synchronization controller.

To improve position synchronization accuracy of the active/active redundant EMA system, the synchronization controller is designed. The schematic of the controller is shown in the Fig. 15. Because of the good position tracking performance of GEMA, the controller makes the GEMA as reference position. DEMA, as a slave, synchronize the GEMA position. The position difference between GEMA and GEMA channel is $\Delta = X_g - X_d$. The difference is used to compensate the position feedback θ_g and θ_d. Then a proportional and integral controller is adopted before the compensation. Avoiding unnecessary and unforeseen overshooting feedback, a saturation output unit is employed, as is shown in the Fig. 15. The X_g and X_d are obtained from the surface angle encoder feedback θ_{cs}. The surface load is set to null.

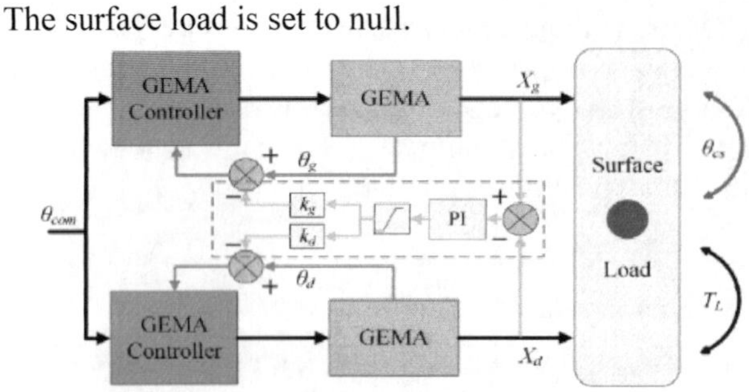

Figure 15. Schematic of synchronization controller.

In this strategy, the position controllers designed for DEMA and GEMA in Fig. 9 are involved, so the only parameter to be valued is the proportional and integral gain for the synchronization controller. The synchronization control strategy should not alter the system performance on pursuit, rejection and stability. Theoretically, the larger the gains value, the faster the synchronization controller being synchronized. But the larger gains will worsen the system stability. So finally, a balance value should be selected. For choosing this value, several experiments are run on test bench. In the end, the proportional gain is 1.45×10-6, and the integral gain is 1.1×10-6/s. The position synchronization tracking curve gotten with this controller parameters is shown in Fig. 16. The 10mm relative position command is given to these two EMAs at the same time. From the Fig. 16, it shows that two EMA channels output almost the same position. There is no apparent error in Fig. 16. In fact, from the plot in Fig. 17, the biggest synchronization tracking error is 0.02mm. More than 30 percent of the error is reduced, compared with no synchronization controller.

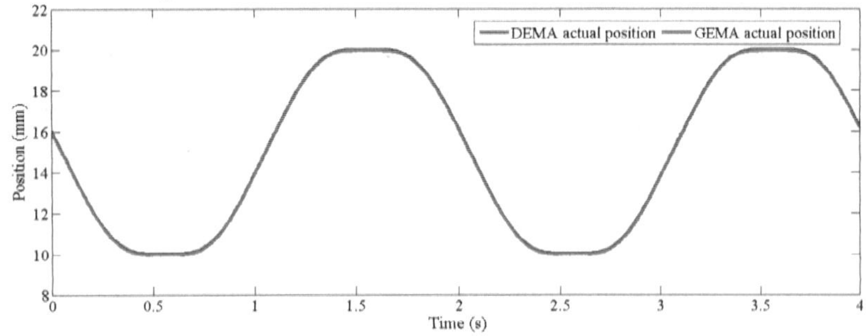

Figure 16. Synchronization tracking curve with synchronization controller.

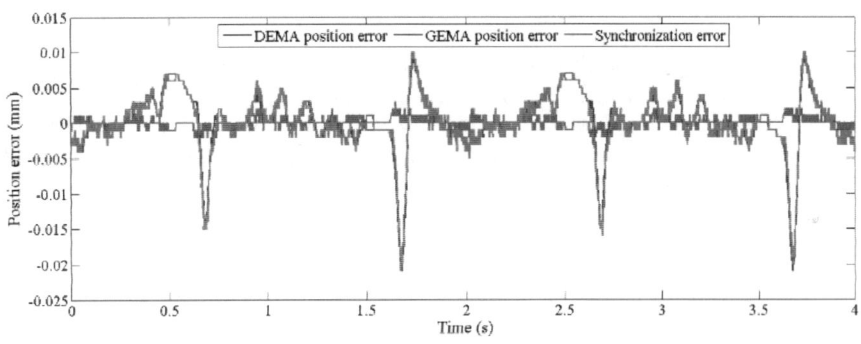

Figure 17. Position tracking errors with synchronization controller.

From the Fig. 14 and Fig. 17, it is easy to find that DEMA has a more tracking error relative to FEMA one. The DEMA deviation is 0.02mm, while GEMA is 0.0025mm, as are shown in the Fig. 17. They are not in the same order of magnitude. The big gap due to different load capacity. It means that FEMA has a bigger stiffness than DMEA. The force output of DEMA have almost no impact on the GEMA's position tracking. In contrast, the DEMA's position tracking accuracy is poor, due to the poor stiffness.

5. Conclusions and future work

In this paper, two EMA position controllers and a synchronization controller are designed, and an active/active redundant electro-mechanical actuation test system is developed to combine the DEMA and GEMA, realizing synchronous motion. A series of synchronous motion experiments are carried out to verify the effectiveness of the design of test system and the synchronization controller. The single channel motion experiments shows that the PID plus position and velocity feed forward controller is very effective for EMA position control. The biggest position tracking errors is 0.1mm. The good performance of these two controller is a firmer foundation for the synchronization controller design. The synchronous motion experimental results show that the synchronization controller can achieve synchronous motion control function well, by decrease more than 30 percent synchronization error.

As future works, the EMA controller and synchronization controller will be verified in other electromechanical system, and a new active/active redundant electro-mechanical actuation test system will be employed as an experimental platform for further study on redundant electro-mechanical actuation system. Meanwhile, the attention will be put on the optimization design of system structure and controllers to further improve the active/active redundant electro-mechanical actuation test system performance. Moreover, influence of the stiffness from inner EMA and rod end mechanical connection is worth researching and solving. The appropriate control method should be adopted to achieve the function requirements for the flight control system. With the improvement of the performance active/active redundant electro-mechanical actuation system, this redundant EMA system can be not only expected to be used in the primary flight control system, but also navigation control, to accomplish aim of the more or all electric aircraft and ship.

Acknowledgments

This study is supported by the Aviation Science Foundation of China (20132851033), International S&T Program of China (2010DFA72540).

References

[1] Wikipedia.org 2014 Wright brothers (http://en.wikipedia.org/wiki/Wright_brothers.Jul.2014)
[2] Song J 2004 Aircraft Architecture (Beijing:Press of Aerospace Industry ISBN 7-80183-336-8) pp 33-39
[3] Wikipedia.org 2014 Aircraft flight control system
[4] William H Philips P and Douglas Arbuckle 2012 Flight controls Accessscience website (http://accessscience.com/content/Flight-controls/260900)
[5] Rosero JA Ortega JA Aldabas E and Romeral L 2007 Moving towards a more electric aircraft vol 22 (Aerospace and Electronic Systems Magazine) chapter 3 pp 3-9
[6] Todeschi M 2010 Airbus – EMAs for flight actuation systems – perspectives (International conference on recent advances in aerospace actuation systems and components) pp 1-8
[7] Moog Inc 2012 System provider for the Boeing 787 Dreamliner (http://www.moog.com/markets/aircraft/civil-aircraft/commercial-transport)
[8] 2012 ACTUATION 2015 (European Commission Framework Programme)
[9] Chenoweth C C and Ryder D R 2000 Redundancy of Hydraulic Flight Control Actuators (Boeing Commercial Airplane Company)
[10] Velinsky S A Lasky T A and Chu B 2009 Kinematics and efficiency analysis of the planetary roller screw mechanism vol 131 (Transactions ASME Journal of Mechanical Design) pp 6-8
[11] Sokolov P A Blinov D S Ryakhovskii O A Ochkasov E E and Drobizheva A Y 2008 Promising rotation-translation converters 28(10) (Russian Engineering Research) pp 949–956
[12] Wang LJ 2012 Force equalization for active/active redundant actuation system involving servo-hydraulic and electromechanical technologies (Toulouse: PhD Thesis INSA) pp 22
[13] Wang X J Wang S P and Wang X D 2009 Electrical Load Simulator Based on Velocity-Loop Compensation and Improved Fuzzy-PID (Seoul:Proceeding of IEEE International Symposium on Industrial Electronics Seoul Korea) pp 238-243

IOP Conf. Series: Materials Science and Engineering **638** (2019) 012008 doi:10.1088/1757-899X/638/1/012008

Sliding Mode Variable Structure Control Strategy Applied on Position Servo System of EMA

D Y Wang[1], S J Gao[1], M Yan[2] and S C Zheng[3]

[1]Department National Key Laboratory of Science and Technology on Aerospace Intelligence Control, Beijing 100854, China
[2] Beijing Institute of Automatic Control Equipment, Beijing 100091, China
[3] Beihang University, Beijing 100074, China

deyiwang@vip.qq.com

Abstract. This paper introduces the basic elements of Electronic-Mechanical Actuator (EMA), and gives the equations. The mathematical model-state equations is established based on the basic equations, and the sliding mode variable structure control strategy, based on the reaching law is applied to control the position servo of EMA. We design a position-speed integrated sliding mode variable structure controller. The models of EMA machinery, electrical motor and controller are established in AMESim and Matlab, and joint simulation is carried out. Finally, this control strategy is compared with the three loop PID control strategy, and the experiment is carried out. The result shows that the sliding mode variable structure control strategy based on reaching law used to control EMA can improve the frequency response and control accuracy of the system, can also improve the dynamic performance of the position servo system.

1. Introduction

As for the small size, light weight and no leakage, EMA will be widely used in aircraft actuating system in the future. However, when EMA is applied on the actuating system of aircraft, the main problems are:

- Surface load condition of the actuation system is not only a function of the rudder angle, but also a complex non-linear function asssociated with many factors, such as air paramiters, aircraft attitud,e Flight Machr and so on, which led to the force condition of EMA changes constantly[1];
- EMA is a multivariable, strongly coupled and nonlinear system;
- Because of the impact resistance of the rolling screw is poor and easy to reverse, the influence of load disturbance must be overcome effectively to ensure the system stiffness [2].

Although the PID control strategy is simple and easy to apply, there are some disadvantages, such as the poor robustness of the time variant perturbation and parameter perturbation, etc.[2], which makes it difficult to solve the three problems above. Therefore, it is an urgent task to adopt a better control strategy to realize the position servo control with high speed, high accuracy and no overshoot.

The sliding mode variable structure control is discontinuous, and the switching characteristic of the system structure with the change of time, which is the essential difference between the sliding mode control and other control strategies. It forces the system to operate along the plan, and the motion with small amplitude and high frequency. So it is called sliding mode motion[3]. The sliding mode can be

managed artificially, and it is independent of the change of system parameters and the degree of external disturbance, which makes the system to be more robust.
The variable sliding mode structure control strategy is applied on EMA position servo control system to improve the system tracking accuracy and dynamic performance and robustness. Through co-simulation analysis of AMESim and MATLAB, the rapidity, frequency response and anti-disturbance ability are compared with tricyclic control strategy. This paper will analyze the merits of the control effect of two control strategies, to verify the correctness and effectiveness of the research results, by experimental verification.

2. Mathematical model of EMA position servo system

EMA is usually composed of servo motor, rolling screw mechanism or gear rotating mechanism. EMA is used to change the rotation motion of the motor into linear motion output. The position servo system of EMA in this paper is a parallel EMA, and a planet roller screw is used as the rolling screw mechanism. The basic structure of the roller screw is shown in "Fig. 1" [4]. It is mainly composed of permanent magnet synchronous motor, planet roller screw rod and other components. Based on the study of the basic components of EMA, the basic equation is established in this paper, which is the basis of the EMA mathematical model.

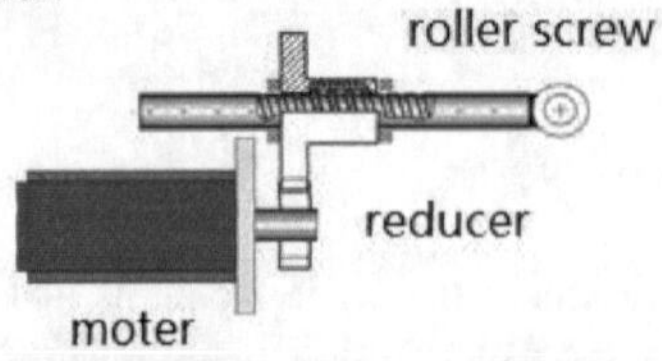

Figure 1. Basic structure of EMA.

2.1. Voltage balancing equation of permanent magnet synchronous torque motor

$$\begin{bmatrix} u_d \\ u_q \end{bmatrix} = \begin{bmatrix} R_s & -\omega_m L_q \\ \omega_m L_d & R_s \end{bmatrix} \begin{bmatrix} i_d \\ i_q \end{bmatrix} + \begin{bmatrix} L_d & 0 \\ 0 & L_q \end{bmatrix} \begin{bmatrix} \frac{di_d}{dt} \\ \frac{di_q}{dt} \end{bmatrix} + \begin{bmatrix} 0 \\ \omega_m \psi_f \end{bmatrix} \quad (1)$$

$$T_m = \frac{3}{2} p_n \psi_f i_q \quad (2)$$

Using the coordinate transformation theory, we establish the model in *d-q* coordinate system[4], controlling i_q with as $i_d=0$, make the electromagnetic torque and current PMSM linearization, the PMSM decoupling control, so this method is simple, it is the most popular [5].
In the formula,
i_q: q axis is current (A); U_q: q is axis voltage; i_d: d is axis current (A); U_d: d is axis voltage (V); Rs: motor stator winding(Ω); L_q: q is axis inductance(H); L_d: d is axis inductance (H); ω_m: motor rotor angular velocity (rad/s); ψ_f :rotor permanent magnet flux psi (Wb).

2.2. Motor end torque balance equation

$$T_m = T_L + B_{fm}\omega_m + J_m\ddot{\theta}_m \quad (3)$$

In the formula, T_L is the external load that applied on the motor torque (Nm); B_{fm} is the equivalent viscous friction coefficient when the motor rotates (rotating resistance coefficient) (Nm s/rad); J_m motor and screw rotational inertia (kg·m²)); θ_m is the mechanical angle of PMSM (rad).

2.3. Torque and force transfer

$$F_n = \eta_{ema} k_{ema} T_L R \quad (4)$$

$$k_{\text{ema}} = 2\pi / P_{\text{rs}} \tag{5}$$

In the formula, F_n: the screw force after motor transformation (N); η_{ema}: transmission efficiency of the roller screw; k_{ema}: roller screw transmission ratio; R: reducer transmission ratio (R=1); P_{rs}: roller screw (mm).

2.4. Force analysis of planetary roller screw

$$F_{\text{n}} = m_{\text{ema}}\ddot{x}_{\text{ema}} + F_{\text{ema}} \tag{6}$$

$$\theta_{\text{m}} = k_{\text{ema}} x_{\text{ema}} \tag{7}$$

In (4), drive efficiency of the roller screw η_{ema} contains viscous friction of the screw, viscous friction of the screw transmission will not be considered[6].

In the model, m_{ema}: roller screw self-mass (Kg); x_{ema}: EMA output linear displacement (m); F_{ema}: external force (N).

2.5. State equation of position servo EMA system

According to (2) - (7), the system input is x_{ema}, ω_{m}, output is i_{qd}, and system disturbance is F_{ema}.

$$x = x_{\text{ema}}, x_1 = x, x_2 = \dot{x} \qquad \ddot{x} = -f(x,t) + bu - Nd(t) \tag{8}$$

In which,

$$f(x,t) = \frac{\eta_{\text{ema}} k_{\text{ema}}{}^2 B_{\text{fm}} R}{m_{\text{ema}}} x_2 + \frac{\eta_{\text{ema}} k_{\text{ema}} R J_{\text{m}}}{m_{\text{ema}}} \dot{\omega}_{\text{m}}$$

$$b = \frac{1.5\eta_{\text{ema}} k_{\text{ema}} R p_{\text{n}} \psi_{\text{f}}}{m_{\text{ema}}};\ u = i_{\text{qd}};\ N = \frac{1}{m_{\text{ema}}};\ d(t) = F_{\text{ema}} \tag{9}$$

3. Design of Position servo controller

The block diagram of position servo controller is as follows:

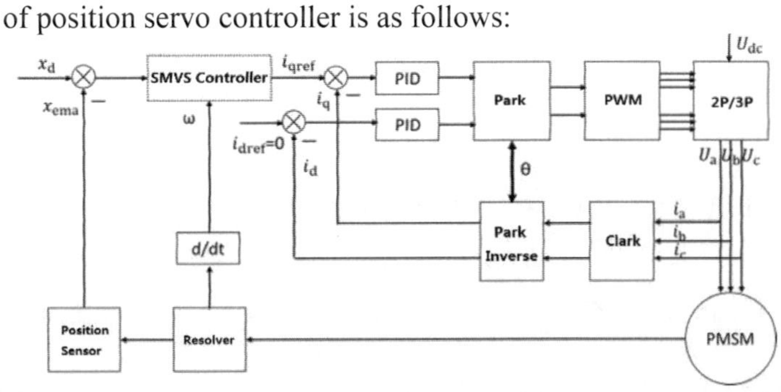

Figure 2. Schematic diagram of position servo integrated sliding mode variable structure controller.

3.1. Design of position and speed integrated sliding mode controller

Using sliding mode variable structure control strategy, an external loop regulator with position loop and speed loop is designed. The input is x_d, position feedback signal x_{ema} and velocity feedback ω. As is known, $x = F_{ema}$, obtained by (8).

3.1.1. Error equation[3] Actual output location $x_{ema} = x$, Set the position control expectation value $x_{emad} = x_d$, then

Definite error:

$$e = x_d - x \tag{10}$$

The corresponding error is divided into

$$\dot{e} = \dot{x}_d - \dot{x} \tag{11}$$

3.1.2. Design of control law The design switching function is

$$s = ce + \dot{e} \tag{12}$$

The upper differential is obtained

$$\dot{s} = c\dot{e} + \ddot{e} = c(\dot{x}_d - \dot{x}) + \ddot{x}_d - \ddot{x} = c(\dot{x}_d - \dot{x}) + \ddot{x}_d + f(x,t) - bu + Nd(t) \tag{13}$$

Take s as exponential approaching law

$$\dot{s} = -ks - \varepsilon sgn(s)\ -ks - \varepsilon sgn(s) = c(\dot{x}_d - \dot{x}) + x_d + f(x,t) - bu + Nd(t) \tag{14}$$

Therefore, the design control rate is

$$u = \tfrac{1}{b}[c(\dot{x}_d - \dot{x}) + \ddot{x}_d + f(x,t) + Nd(t) + ks + \varepsilon sgn(s)] \tag{15}$$

3.1.3. Proof of stability
Lyapunov function

$$V = \tfrac{1}{2}s^2 \tag{16}$$

For the upper differential

$$\dot{V} = s\dot{s} = s(c\dot{e} + \ddot{e}) = s[c(\dot{x}_d - \dot{x}) + \ddot{x}_d - \ddot{x}] = s[-ks - \varepsilon sgn(s)] = -ks^2 - \varepsilon|s| \le 0 \tag{17}$$

The control law is proved to be asymptotically stable in Lyapunov sense, namely global convergence.

3.2. Design of current loop PID controller
The current ring is a very important part of the AC motor position servo control system. The current loop can improve the fast response of the system and suppress the internal interference. In this paper, the current loop adopts the $i_d=0$ control method and uses the PI control to improve the system stability while taking into account the system precision.

4. Construction of EMA joint simulation platform

EMA is consisted of two parts, mechanical part and control part. In AMESim, EMA is conducive to the establishment of a more accurate mechanical model[7], MATLAB has a big advantage in the control system modeling[8], so the mechanical model established in AMESim as shown in "Fig.3", the control model set up in the MATLAB/Simulink is as shown in "Fig.4".
In fact that, the speed, current, voltage and other signals cannot be infinitely large, so certain restrictions need to make for them.
Where the maximum amplitude of the phase current allowed by the motor is expressed in I_{max}, the current limiting condition is:

$$I_d^2 + I_q^2 \le I_{max}^2 \tag{18}$$

When the maximum phase voltage is indicated by V_{max}, the voltage limiting condition is [9]:

$$\sqrt{U_d^2 + U_q^2} \le V_{dc} / \sqrt{3} \tag{19}$$

Finally, the nonlinear factors such as system current limiting and voltage limiting is added into the model of the system control section,

The interface is provided by AMESim, and the mechanical part outputs F_{ema}, T_L and x_{ema}, which are received by S-Function in MATLAB/Simulink and input into the control system. In this control system, the sliding mode variable structure control strategy is written as the m file, control current i_{qd} is obtained by the signal processing, i_{qd} controls the motor by PI control, output speed signal. Output to AMESim through S-Function, controls the mechanical system, so that the output position is equal to the command position signal.

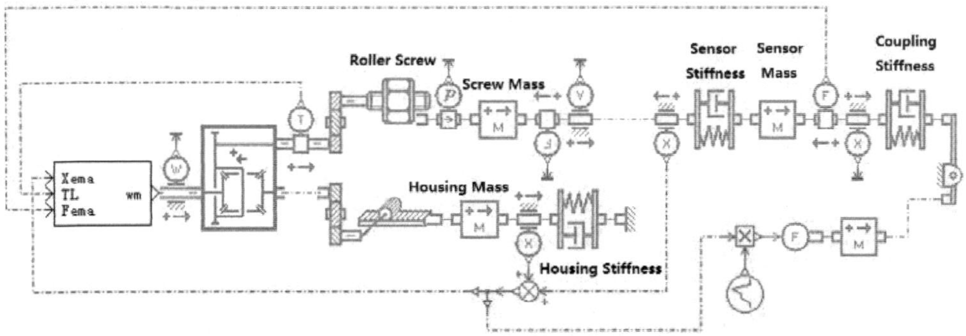

Figure 3. AMESim model of EMA mechanical part based on sliding mode variable structure.

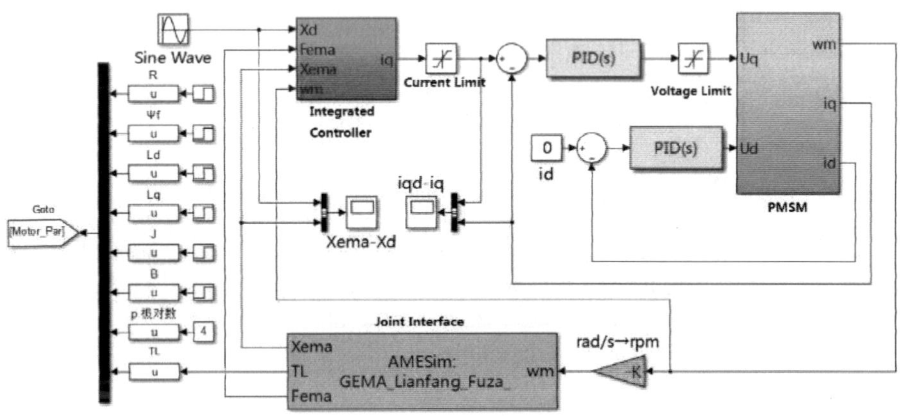

Figure 4. MATLAB model of EMA control part based on sliding mode variable structure.

5. Simulation results and Analysis

Simulation parameters are shown in Table I.

Table 1. Ema system parameters.

Index	Value	Index	Value
L_s	0.0227H	k_{ema}	1256.6
R_s	3.85Ω	m_{ema}	15kg
B_{fm}	0.00473N·m·s/rad	p_n	4
ψ_f	0.2381Wb	m_l	15kg
P_{rs}	5mm	η_{ema}	90%
J_m	0.0013kg·m^2		

Tricyclic PID control strategy is compared with sliding mode control strategy, based on step response, sinusoidal response and anti-jamming capability.

5.1. *Compare the indexes of the two different control strategies*

Table 2. Step response comparison.

Index	Sliding mode control strategy	Tricyclic PID control strategy
risetimet_r/s	0.6397	0. 6405

regulation time t_s/s	0.7686	0.7742
overshoot /%	0.9	0.23
steady state error /%	0.05	0.15

As shown in table 2, the control effect is similar, from the comparison of rise time, regulation time, overshoot and steady state error of the two control strategies. Sliding mode variable structure control strategy is not obvious, even through overshoot is slightly higher than three ring PID.

The step response of sliding mode control strategy and tricyclic PID control strategy is shown in “Fig. 5”.

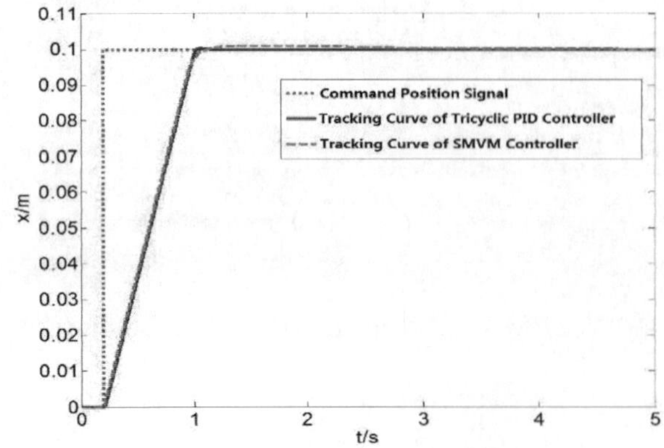

Figure 5. Step response (position command =0.1m).

5.2. *Compare the sinusoidal response's rapidity precision of the two control strategies*

Table 3. Sinusoidal response comparison.

Frequency/Hz	Tricyclic PID control strategy		Sliding mode control strategy	
	Amplitude attenuation /%	Phase lag /%	Amplitude attenuation /%	Phase lag°
1	0.13	6.9	0.08	3.3
3	1.64	21.2	0.8	11.2
5	8.5	40.3	3.8	34.6
7	29	62.2	29	58.2
7.1	30	63.4	—	—
7.5	—	—	29.28	69.2

From the response curve of the two control strategies, The limitation of the system's sinusoidal frequency response is not the current limit, but the speed limit. When the system is designed, the system speed limit is slightly increased, and the system frequency response can be enhanced.
Two kinds of control strategy, sinusoidal response, amplitude attenuation, phase lag and shock degree were compared as shown in Table 3, sliding mode variable structure control in system response and tracking accuracy (amplitude attenuation and phase lag degree) were slightly better than tricyclic PID control strategy.

5.3. *Compare the two control strategies against external disturbances*

When the two control strategies under the conditions of "0s, 5000N/s ramp force" and "2S, 10kN force", the comparative analysis of the output position is shown in “Fig. 6”.

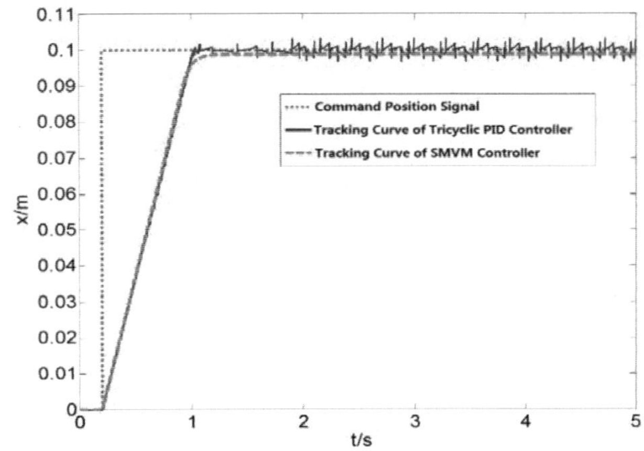

Figure 6. Comparison of the two control strategies of position servo under the condition 1.

When the two control strategies under the conditions of "2s, adding 10kN step force" and "3s, unloading", the comparative analysis of the output position is shown in "Fig. 7".

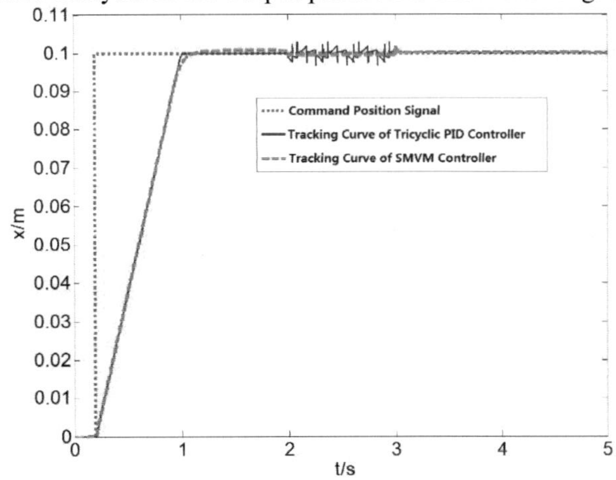

Figure 7. Comparison of the two control strategies of position servo under the condition 2.

When the two control strategies under the conditions of "10kN/50mm elastic load", the comparative analysis of the output position is shown in "Fig. 8".

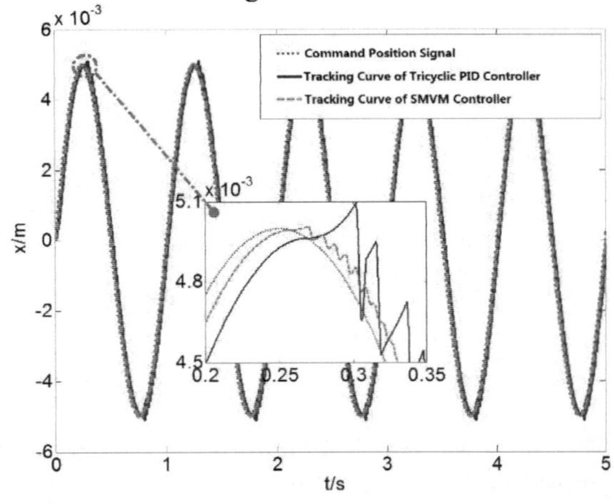

Figure 8. Comparison of the two control strategies of position servo under the condition 3.

By comparing the output of position servo curves above three conditions, ring PID control strategy to control the output of the system, the external disturbance situation, there will be varying degrees of jitter, and the sliding mode variable structure control system control strategy, anti disturbance ability was stronger than the tricyclic PID control strategy, robustness.

6. Experimental verification

The experimental verification is carried out. The contrast curve between the instruction position signal, the simulation results and the experimental results are shown in "Fig. 9".

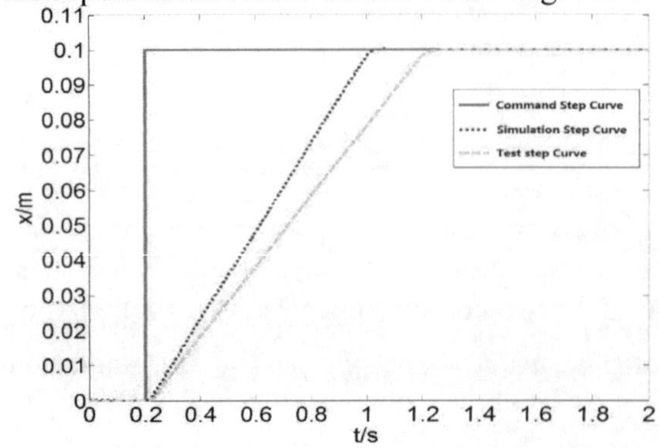

Figure 9. Comparison curves between experiment and simulation results.

The experimental results show that the control strategy is correct, effective and feasible.

7. Conclusions

The sliding mode variable structure design is used on EMA position servo system controller, and realize the position control of EMA. Through AMESim and MATLAB/Simulink co-simulation, and compared with ring PID control strategy, we get some experiment results. The results show that the sliding mode variable structure control strategy can improve the system dynamic performance and position accuracy, which ensures the robustness of the system in the external interference. The results show that it is an effective control method.

Acknowledgments
This study is supported by the Aviation Science Foundation of China (20132851033).

References

[1] Wang C 2010 Design and research of electric servo load simulator (Harbin Institute of Technology)

[2] Fu Y Liu H and Pang R 2011 Research on Active Disturbance Rejection Control for Direct Drive Electromechanical Actuator (Journal of Sichuan University Engineering Science Edition) vol 43 pp 121-127

[3] Liu J 2005 *Sliding Mode Variable Structure Control MATLAB Simulation (Beijing)* (Tsinghua University Press) p 154-167

[4] Wang J 2015 *Permanent Magnet Synchronous Motor Intelligent Control Technology* (Southwest Jiaotong University Press) p 97-130

[5] Kou B 2008 *AC servo system and its control* (Mechanical Industry Press) p 214-235

[6] Qiao G Liu G and Ma S 2008 *Dynamic Characteristics Analysis of Electromechanical Actuator Based on Planetary Roller Screw Pair* vol 35 (Journal of Vibration and Shock) pp 82-101

[7] Fu Y Qi H 2011 *LMS imagine. Lab AMESim System Modeling and Simulation Example Tutorial* (Beijing) (Beijing Aerospace University Press) p 89

[8] Shi L 2014 *MATLAB/Simulink System Simulation Super Learning Manual* (Beijing) (People's Posts and Telecommunications Press) p 200-231

[9] Zhu L Wen X and Zhao F 2011 *Research on Weak Magnetic Loss Control Mechanism of Permanent Magnet Synchronous Motor and Its Countermeasures* vol 31 (Proceedings of the CSEE) pp 67-72

IOP Conf. Series: Materials Science and Engineering **638** (2019) 012009 doi:10.1088/1757-899X/638/1/012009

Research on automatic assembly technology for final assembly of helicopter fuselage

Q D Xiao[1], L A Zhang[2], L Gan[1] and T Cheng[1]

[1] AVIC Manufacturing Technology Institute, Beijing, 100024, China
[2] AVIC Hafei Aviation Industry Co., Ltd, Harbin, 150060, China

Abstract. The helicopter fuselage has the characteristics of compact structure space, complex coordination relationship and weak rigidity, etc. It is difficult to control the accuracy of the desired shape. According to the high precision and quality requirement of final assembly for helicopter fuselage, the application status of automatic assembly technology in aviation manufacturing was analyzed. Then the difficulties of automatic assembly technology for fuselage final assembly were also discussed. In this paper, an automatic assembly system for the final assembly of helicopter fuselage was designed, and relevant key technologies of the assembly system mentioned above were proposed. The designed system have the functions of automatic positioning, clamping, measuring, drilling, feeding and blind fastener riveting for fuselage components, which can improve the assembly quality and efficiency of helicopter fuselage during the final assembly process.

1. Introduction

In aircraft manufacturing, due to the large size, complex shape, and the large amount of parts and fasteners, the workload of aircraft assembly accounts for about half or more of the total workload [1,2]. Automatic assembly technology represents for the development trend of aircraft manufacturing technology, and it is also one of the key solutions to improve the quality and efficiency of aircraft assembly and to ensure the long life of aircraft [3,4]. In the aviation enterprises of China, to the fixed-wing aircraft manufacturers, they have developed the research and application of automatic assembly technology and equipment, such as (1) automatic drilling[5,6], (2) automatic riveting of panels[7-9], (3) automatic assembly of components[10,11], (4) automatic docking of large components [12] and so on. Compared with fixed-wing aircraft, helicopter assembly still adopts traditional manual assembly methods, i.e. the positioning and clamping components with a large number of complex rigid toolings. Automatic assembly, which having a huge influence on the assembly quality and efficiency of helicopter fuselage, is basically in a blank state.

Due to the internal structure of helicopter is compact, and the coordination relationship is very complicated. During the assembly process of fuselage structure, the positioning and clamping of product components, drilling and fastening, especially for the riveting of blind fasteners, are the key factors to ensure the assembly quality and efficiency. In the actual assembly site, the accuracy of positioning and clamping has a direct relationship with the position accuracy and shape accuracy of fuselage structure. And the quality of drilling affects the long life of the fuselage. In addition, the amount of blind fasteners is huge (for example, there are about 12000 blind fasteners in AC352 helicopter fuselage structure). Some problems, such as unstable quality and low efficiency, often occurred in manual drilling and riveting.

Aiming at the existing problems in the final assembly process for helicopter fuselage, firstly, the key technologies of automatic assembly of helicopter fuselage final assembly were studied, and then an automatic assembly system for final assembly of helicopter fuselage was designed. The system has the function of automatic positioning, clamping, measuring, drilling, feeding and fastener riveting of fuselage components. Finally, the assembly quality and efficiency of helicopter fuselage final assembly was improved.

2. Application status of automatic assembly technology

Automatic assembly technology is widely used in the field of foreign aviation manufacturing, especially in the assembly of large fixed-wing aircraft. B & C company, Italy, provided automatic assembly lines to Boeing and Airbus for the assembly of B787 and A380 fuselage sections. Four automatic assembly lines, which including robotic riveting system, are constructed for the 44th and 46th sections of B787 Dreamliner. As shown in figure 1, the robotic riveting system includes two sets of robotic systems, i.e. external robotic system and internal robotic system. The two systems work together to achieve the automatic drilling and the lock bolt riveting of composite fuselage panel. It is mentioned that, the external robotic system is responsible for automatic measuring, clamping, posture adjustment, drilling and countersinking, absorption, sealant application and bolt inserting, while the internal robot system is responsible for automatic clamping, absorption, collar installation, riveting, and etc.

Figure 1. Robot automatic riveting system of B787 fuselage section.

Comparing with the aircraft with fixed-wing, helicopter airframe has smaller size and tight operating space. As a result, it is very difficult to apply automatic assembly equipment or tooling, because they often have a large size. In recent years, with the development of manufacturing technology, the automatic and digital assembly technology in the whole industry, is becoming more and more mature for helicopters. Relevant technologies and equipment are developed in some famous aircraft companies, such as Airbus, Bell and Sikorsky, and they have been widely used in the assembly process of helicopter body parts and subsystems.

As shown in figure 2, the robot automatic assembly system of a helicopter fuselage assembly in Airbus Helicopter Company is introduced. The system clamps and locates the fuselage structure with a set of automatic positioning tooling. The automatic assembly of the new helicopter fuselage structure is developed by digital measurement, precise positioning of the robot, automatic drilling and riveting and other technologies. With the system, the total assembly time is less than 28 hours. And the system can improve the assembly quality and efficiency of fuselage structure greatly. Obviously, the high performance demands of helicopter structure can be fulfilled.

Figure 2. Automatic assembly system for a helicopter.

3. Technical Characteristics of Automatic Assembly System for Final Assembly

The structural characteristics for final assembly of helicopter fuselage can be summarized as follows: 1) the fuselage has a cylindrical structure, and the assembly working area has a circumferential distribution; 2) the fuselage also has a weak rigidity, correspondingly, it requires maintaining the required shape that defined by the official assembly documents in the assembly process. While, to the riveting system that working with an automatic robot, it has the following characteristics, as shown in figure 3.

Firstly, with the view of structural form, it carries the end-effector to work.

Secondly, due to the structure limitations, the working scope of the robot is limited in a certain space comparing with the whole fuselage.

Thirdly, the positioning accuracy of the robot is not high enough.

Figure 3. Automatic assembly system for final assembly of helicopter fuselage.

Due to the characteristics of the above technologies, in this paper, the automatic assembly system for final assembly of helicopter fuselage is mainly composed of (1) rotary positioning tooling system, and (2) robot automatic riveting system. To the robot riveting system, it includes three main parts, i.e. 1) the 7th axis mobile platform, 2) the robot, and 3) the end-effector. The above system is mounted on the 7th axis mobile platform with the end-effector to be a whole tooling unit. With the linear motion of the 7th axis, and the rotation driven by the rotary tooling, the working limitation of the robot can be compensated effectively.

There are four hinge joints for main reducer on the fuselage, with which the rotary positioning tooling system locating and clamping the helicopter fuselage. Consequently, during the rotating process, the tooling system positions the fuselage structure, and the pneumatic shape was also controlled to the desired requirement. It is mentioned that the working space of robot riveting system is limited, which can only cover a quarter of turn the fuselage structure. However, with the help of rotary tooling, the remaining there quarters of assembly area for the fuselage structure can be completed. To the rotating tooling system, it carries the fuselage body component rotating 90 degree at one time. And with the help of the robot riveting system, the riveting of the current assembly area can be accomplished. In other words, the assembly of the whole body circumference can be completed for four rotations of the tooling system.

Based on the structure of the assembly system, the main work flow of the system can be described as follows.

1) After the pre-assembly work of front fuselage, mid fuselage and transitional section is completed, the digital rotating flexible tooling is designed to locate and clamp the fuselage structure automatically.
2) The robot carrying the end-effector moves linearly through the 7th mobile platform, to the position of the riveting station.
3) Then with the help of photographic measurement unit mounting on the end-effector, the datum holes in the fuselage structure can be measured, and coordinate system between the workpiece and robot riveting system can be kept in a harmonious status.
4) The robot riveting system moves to the first assembly point, to perform the operations including automatic drilling, feeding and blind fasteners riveting.
5) After that, the system moves to the next point to drill and rivet, and recycles this operation until the assembly operations are completed in the concerning area.
6) The rotary tooling system drives the body structure to rotate, and then robot riveting system starts the following assembly work, and it will recycle it until the assembly operation of the whole body is completed.

4. Key Technology during Automatic Assembly

Automatic assembly technology of helicopter fuselage final assembly covers tooling technology, robotics technology, measuring technology, drilling and riveting technology, and integrated control technology, etc. It covers a wide range of majors, and it is also difficult to achieve. The following key technologies need to break through.

4.1. Automatic positioning for fuselage final assembly

According to the characteristics of helicopter fuselage structure, the positioning modes of linear motion tooling and circumferential motion tooling are compared and analyzed, and a form of rotating automatic positioning tooling with circumferential motion is defined. As shown in figure 4, the tooling includes the main frame of the tooling, rotating mechanism, modular positioner and modular sustaining mechanism, so as to complete the automatic rotating, positioning and shape controlling of helicopter fuselage section.

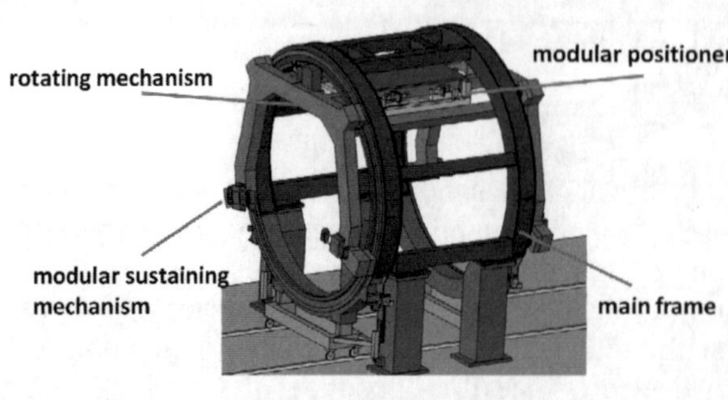

Figure 4. Rotary positioning tooling for fuselage structure.

In order to realize the automatic positioning of helicopter fuselage structure, the structure needs to be designed for the automatic positioning[13]: 1) In order to enhance the rigidity of the fuselage structure at the breakdown interface and reduce the deformation of the fuselage during the positioning process, the production breakdown interface is necessary to be optimal designed[14,15]. 2) To facilitate the rapid connection between the fuselage structure and the fixture, and shorten the installation and position time of the fuselage, the hinge joints of the fuselage, which matching with the modular positioner of rotary tooling, also need to be designed. 3) To the design and layout of the measuring datum points (holes), they should be set in the rigid area of the fuselage as far as possible and distributed uniformly, and then they would envelop the fuselage in a wide range so as to improve the positioning accuracy of the riveting system and products.

The main frame of the tooling is consisted of four basing pillars, and a set of annular structure. To facilitate product to enter and exit with good convenience, the annular device is divided into two arc portions, i.e. the upper and lower. The rotating mechanism includes the main structure and the driving mechanism. The driving motors and gears are installed together with the main structure. Because the double drives move synchronously, so it can drive the product rolling. To the modular positioner, it includes base mechanism, lifting drives, positioning and connecting structure, process joints, and other parts. The positioner could connect with the fuselage structure quickly to lift and position the product. Besides, the modular sustaining mechanism is joined with the product to help control the shape of the body during the rotating motion.

After the pre-assembly optimization of the integral helicopter fuselage is completed, the fuselage structure then is carried to a required assembly position, i.e. the position that connecting the main hinge joints of the fuselage, with a special carriage. After the modular positioner is joined with the fuselage structure, then the product can be lift and clamped to the desired position. It is mentioned that the modular sustaining mechanism clamps the fuselage to maintain the shape of the aircraft. And the rotating mechanism drives the fuselage to rotate until to the working position, then the assembly operations of the robot automatic riveting system can be carried out.

4.2. End-effector design for automatic drilling and blind fastener riveting

For composite structure airframe, drilling and blind fastener riveting are the key assembly process. The technology of end-effector of automatic drilling and blind fastener riveting is matured abroad[16], and has been applied in the assembly of helicopter fuselage, composite wing, and other components. At present, there are some preliminary applications of automatic drilling in China, but the automation of blind fastener riveting is still in the research status, and has not been applied in aircraft manufacture.

To the end-effector technology which integrates automatic drilling and blind fastener riveting, there are many difficulties. Such as, 1) the structure of the end-effector is complex, and the transposition form of each functional unit is hard to determined; 2) the blind fastener has a slender structure, which makes it difficult to design the structure of automatic feeding unit; 3) the design of the blind fastener riveting unit is compact, due to the narrow space of the product; 4) the integration and control of the end-effector with the clamping, photographic measurement, drilling, feeding, blind fastener riveting, absorption, posture adjustment and other functional units.

As shown in figure 5, the end-effector of automatic drilling and blind fastener riveting includes photographic measurement unit, clamping mechanism, automatic drilling unit, feeding unit, automatic sealing unit, blind bolt riveting unit, posture adjusting unit and etc. It has the functions of automatic positioning with product's feature, posture adjustment, clamping, drilling and countersinking, feeding, automatic sealing, riveting, etc., which can assemble the helicopter fuselage structure automatically.

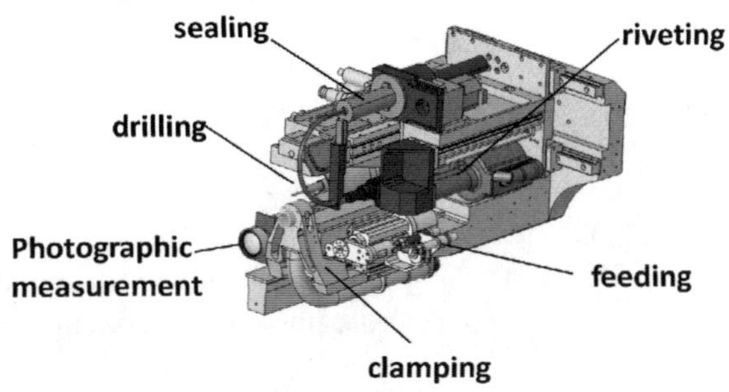

Figure 5. End-effector technology of automatic drilling and blind fastener riveting.

4.3. Precise adjustment of position and posture of automatic riveting system with digital measurement

Accurate control and adjustment of robot riveting position and pose based on digital measurement technology, is the important solutions of high precision assembly of helicopter fuselage structure. The main technical difficulties can be summarized as follows: 1) the establishment of the relative position relationship of fuselage structure, tooling and robot, with the help of photogrammetric measurement, and precise adjustment of the holes' position. 2) The accurate adjustment of the position and the pose of the robot, based on skin surface vector measurement of drilling and riveting.

4.3.1. Accurate positioning technology of robot automatic riveting base on on-site measurement. The datum holes in the fuselage structure are photographically measured and the coordinate system of the fuselage is aligned to determine the actual position of the robot automatic riveting. Its working process can be described as follows: firstly, the control system sends image acquisition command through Ethernet, and then opens the light source. Secondly, the CCD camera begins to take the image of the datum holes of the part. By calculating and analyzing the image of the datum holes, the information of the datum hole is obtained. With the visual algorithm, the position deviation between the axis of the positioning hole and the theoretical coordinate can be calculated, and then the data would be fed back to the control system. Therefore, the automatic riveting system is positioned for another time to ensure the holes' vector before drilling. The above principle can be explanation with figure 6.

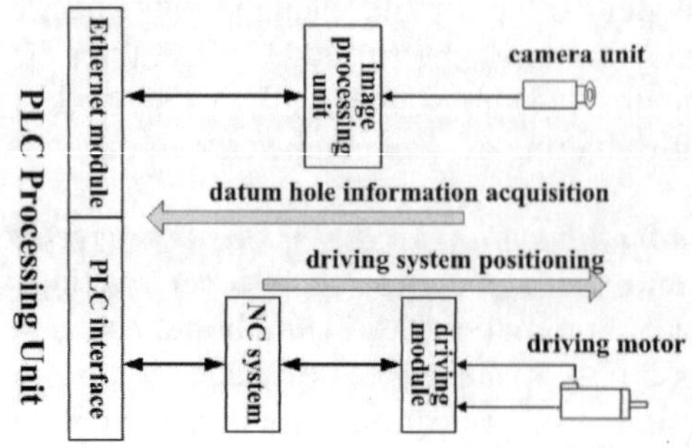

Figure 6. Principle of photographic measurement unit.

4.3.2. Accurate adjustment technology of automatic riveting posture base on laser measurement of surface vector. The multi-point laser rangefinders are used to calculate the vector of the riveting point synchronously, and then the posture of the riveting end-effector is adjusted. As shown in figure 7, four laser rangefinders are designed and installed in the clamping mechanism of the end-effector in this system. The rangefinder emits laser and receives reflected signal, then calculates the linear distance synchronously. According to the distance data of the four rangefinders, the vector of the skin at the holes' position can be approximately fitted. The automatic riveting system adjusts the drilling and riveting posture accurately according to the vector data.

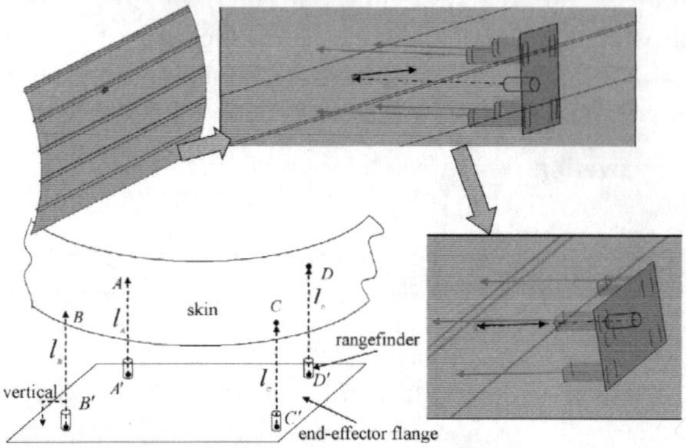

Figure 7. Surface vector measurement and accurate posture adjustment.

4.4. Multi-system integration for automatic assembly of fuselage structure

Data from off-line programming system is imported by computer integrated control[17,18]. According to process parameters, the cycle control of the end-effector and the positioning control of the drilling robot can be executed as planned. In detail, to fulfill the requirements of automatic assembly, the cycle control contains coordinated control of feeding system, photographic measurement of datum hole, on-site visual surveillance and so on.

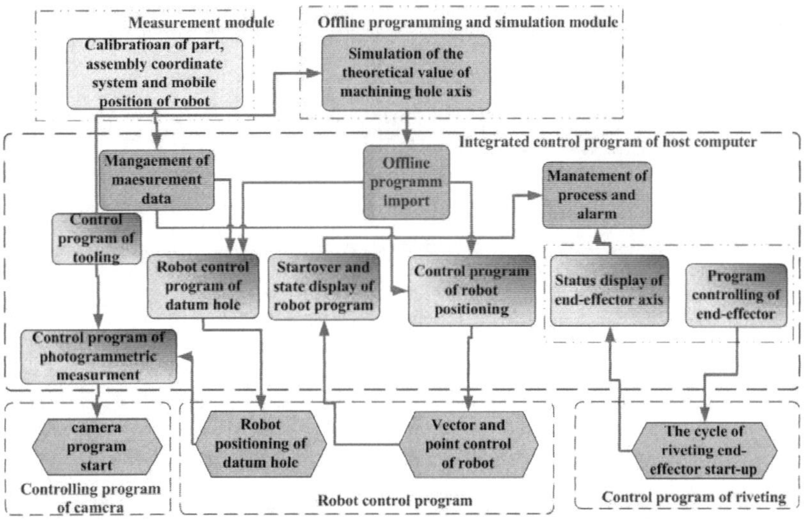

Figure 8. Multi-system coordination control chart.

There are many subsystems in the control system of automatic assembly system. While, there is also measurement information. During the above integration process, the KRC4 controller of the robot itself is used for the robot control, the control module of Kollmorgen is used for the rotary tooling control, and the end-effector is controlled by the other CNC system. It is necessary to build a platform to achieve the information exchange and coordinated control between different working systems. The system of the paper uses integrated control software of the host computer to achieve the functions above. The integrated control software of host computer communicates with other units in the system through the form of Ethernet communication, and then the control and sequence, measurement and security protection of each sub-unit can be in a harmony status. As shown in figure 8, by the feedback of various information, the host computer judges whether the robot and the end-effector are running according to the orders sent from itself or not, and whether they are in the correct place or not. The host computer also manages the data such as measurement data, process parameters, and alarm records, and so on.

The automatic assembly system of fuselage final assembly is controlled by integrated software. The integrated software coordinates the control sequence of each system according to the process flow. The software receives the data of drilling and riveting position from off-line programming system, and then with which it completes the cycle control of drilling and riveting, and manages and records the parameters, as shown in figure 9.

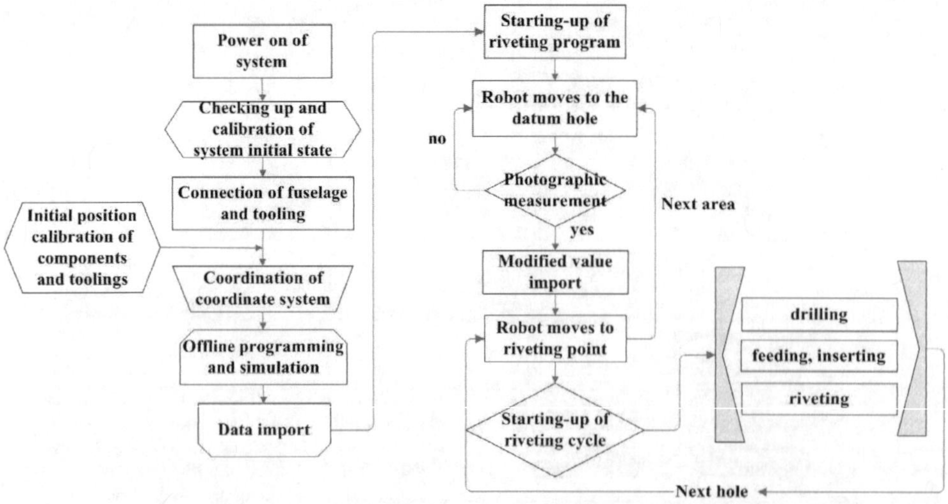

Figure 9. Integrated control flow of automatic assembly system.

5. Conclusion

According to the characteristics of helicopter fuselage structure and the requirement of assembly with high precision and high quality, relevant key technologies, such as, (1) automatic positioning tooling for helicopter fuselage final assembly, (2) automatic riveting end-effector for blind fasteners, (3) precise adjustment of position and posture during riveting process, and (4) multi-system integration are studied. Finally, the automatic assembly system for final assembly of helicopter fuselage is designed in this paper.

To the automatic assembly system for final assembly of helicopter fuselage, it has the characteristics of flexibility, digitalization and modularization, and which can complete the function of automatic positioning, clamping, measuring, drilling, feeding and riveting of blind fasteners of fuselage components and so on. In order to achieve the engineering application of the system in the production of helicopter fuselage final assembly as soon as possible, a large number of process tests should be done to optimize and improve the maturity of the system.

6. References

[1] Zou J H, Riaz A H and Fan Y Q 2006 Research for major-parts digital assembly system of large-scale airplane *Wseas International Conference on Circuits* pp 337–43

[2] Wu D W and Yuan J C 2012 Based on flexible automation fixture of siding components of assembly craft planning and simulation *Applied Mechanics and Materials* **224** pp109-12

[3] Sun W 2013 A laser tracking-vision guiding measurement system for large-scale parts assembly *Advanced Materials Research* **718-720** pp 868-74

[4] Gitta M 2013 *Destination Myachkovo Airport* (Los Angeles: Sent Publishing)

[5] Jin L, Shi X and Dong H Y 2014 Study on robot automatic drilling of carbon fiber reinforced plastics (CFRP) *Advanced Materials Research* **889-890** pp1144-9

[6] Bi Y B, Li Y C and Jiang Y H 2013 An industrial robot based drilling system for aircraft structures *Applied Mechanics and Materials* **433-435** p7

[7] Lin H C 2007 An internet-based automatic riveting system using remote monitoring and control *IEEE International Conference on Systems*

[8] Feng J 2017 Design of the control system for automatic riveting machine based on PLC *International Conference on Robots & Intelligent System*
[9] Jiang L P, Chen W L and Wang M 2014 Approach to interference aircraft automatic riveting process control of drilling and riveting *Transactions of Nanjing University of Aeronautics & Astronautics* **31(6)** pp 609-14
[10] Zhen M and Jing 2011 Automatic assembly for combined mold components based on SolidWorks *International Conference on Electronics*
[11] Zhang H, Guo H J and Zhang S S 2014 Manufacture of flexible automatic assembly platform for aircraft component level *Applied Mechanics & Materials* **635-637** pp 1857-60
[12] Lang Y D, Yao Y X and Xia P J 2008 Research on exact placement technology of virtual assembly for large-scale products *Applied Mechanics & Materials* **10-12** pp 460-5
[13] Shpitalni M, Elber G and Lenz E 1989 Automatic assembly of three-dimensional structures via connectivity graphs *CIRP Annals - Manufacturing Technology* **38 (1)** pp 25-8
[14] Jin Y, Abella R and Ares E 2013 Modeling and digital tool development of a new similarity metric for aerospace production *International journal of advanced manufacturing technology* **69(10)** pp 1-4
[15] Padillo A, Racero J, Oliva M and Mas F 2017 Design and implementation of a prototype for information exchange in digital manufacturing processes in aerospace industry *Product lifecycle management*
[16] Bullen and George N 2013 Automated/mechanized drilling and countersinking of airframes (Warrendale: SAE International Press)
[17] Aehnelt M and Bader S 2014 Tracking assembly processes and providing assistance in smart factories *Icaart* pp 161-8
[18] Campos A, Johnson R and Kennedy J 2016 System and method for wiring an aircraft (US Patent: 2016004863)

Acknowledgements

The authors gratefully acknowledge the support of Ministry of Industry and Information Technology of China (No. MJ-2015-*), and National Defense Industrial Technology Development Program of China (No. JCKY2018205A001).

IOP Conf. Series: Materials Science and Engineering **638** (2019) 012010 doi:10.1088/1757-899X/638/1/012010

Study on abnormal behavior of human in safety accidents of special equipment

Hongqi Luo[1] and Xia Liu[2*]

1 School of Material and Mechanical Engineering, Beijing Technology and Business University, Beijing, China

2 Quality Management Branch of China National Institute of standardization, Beijing, China

E-mail: lhqfrq@126.com

*Corresponding author

Abstract. The data of the total number of accidents and the number of accidents caused by illegal use of special equipment in China was obtained. The prediction models for the safety accidents of special equipment are establish with grey theory based on the actual data. Then the models are processed through optimization. The results show that the simulation values fit well with the real values, which can be used as a theoretical basis and reference for special equipment safety policy formulation and scientific regulatory decision-making.

1. Introduction

Special equipment is an important driving force and tool to ensure the economic and social development, as well as an important infrastructure to ensure the quality of life of people. Because of the large number of people contacted or affected in the whole life process of special equipment, its structural characteristics are dangerous, and the medium carried by some equipment is harmful. Therefore, the safety of special equipment is directly related to the safety of people's lives and property. The safety situation of special equipment is not optimistic in China. The level of various accident indicators is 4 to 6 times that of developed countries. Serious and extraordinary accidents still occur from time to time, and the safety situation has not yet been fundamentally changed [1, 2].

The abnormal behavior of relevant personnel is the main reason for the accidents of special equipment. In order to improve the safety status of special equipment and meet the needs of the development of special equipment, it is significance to analyze the characteristics and rules of the development of special equipment safety accidents for the safety supervision.

2. Data collection and collation

Through consulting "China Heavy Machinery Industry Yearbook" [3], the actual data obtained and the data processed are shown in Table 1. Because of the randomness of the original data column of the number of accidents, it cannot be directly used for modeling, so the original data column is transformed by moving average, such as moving average processing of the statistical data from 2008 to 2016 [1]. The formulas are shown as follow, which t is from 1 to N.

$$X^{(0)}(t)=\left[X^{(0)}(t-1)+2X^{(0)}(t)+X^{(0)}(t+1)\right]/4$$

$$X^{(0)}(1)=\left[3X^{(0)}(1)+X^{(0)}(2)\right]/4$$

$$X^{(0)}(N)=\left[X^{(0)}(N-1)+3X^{(0)}(N)\right]/4$$

The abnormal behavior of relevant personnel is the main reason for the accident of special equipment. The abnormal behaviors of people include the use of illegal equipment, improper maintenance of safety accessories, illegal operation of operators and other illegal use. It shows that the proportion of accidents caused by illegal use in the total number of accidents is 37.80%~65.67% from Table 1. It is very important to study the abnormal behavior of human beings, which is reflected in the safety accidents of special equipment.

3. Prediction models of GM (1, 1)

The grey forecasting method is to establish a mathematical model from the past to the future based on the known or unascertained information of the past and the present, so as to determine the trend of the development of the system and provide a theoretical basis for planning and decision-making [1, 2]. GM (1, 1) is a first-order and variable differential equation model, which is suitable for predicting the development and change of eigenvalues of system behavior [4-6].

Table 1. Number of illegal use and total accidents of special equipment in 2008-2016

Year	Number of accidents		Number of accidents caused by illegal use		Proportion of caused by illegal use
	Actual value	Processed value	Actual value	Processed value	
2008	307	325.25	131	134.25	42.67
2009	380	340.75	144	141.25	37.89
2010	296	311.75	146	137.25	49.32
2011	275	268.50	113	125.75	41.09
2012	228	239.50	131	124.25	57.46
2013	227	241.25	122	121.75	53.74
2014	283	262.50	112	119.00	39.58
2015	257	257.50	130	131.25	50.58
2016	233	239.00	153	147.25	65.67

3.1. Prediction model of accident number Caused by illegal use

The original discrete data sequence of the number of accidents caused by illegal use.

$$x^{(0)} = \left\{x_1^{(0)},\ x_2^{(0)},\ \dots,\ x_N^{(0)}\right\}$$

Where N is the sequence length, N = 9.

$$x^{(0)} = \{131,\ 144,\ 146,\ 113,\ 131,\ 122,\ 112,\ 130,\ 153\}$$

After moving average transformation, the following results are obtained:

$$x^{(0)} = \{134.25,\ 141.25,\ 137.25,\ 125.75,\ 124.25,\ 121.75,\ 119.00,\ 131.25,\ 147.25\}$$

The superscript (0) denotes the cumulative number of times.

$$x_k^1 = \sum_{j=1}^{k} x_j^{(0)},\quad k = 1,\ 2,\ 3,\ \dots,\ N \tag{1}$$

The sequence can be obtained by a process of cumulative generation when it is calculated with formula (1).

$$x^{(1)} = \{134.25,\ 275.50,\ 412.75,\ 538.50,\ 662\,75,\ 784.50,\ 903.50,\ 1034.75,\ 1182.0\}$$

Data matrix B and yN can be obtained as below.

$$B = \begin{bmatrix} -\left(x_2^{(1)} + x_1^{(1)}\right)/2, & 1 \\ -\left(x_3^{(1)} + x_2^{(1)}\right)/2, & 1 \\ \cdots & \cdots \\ -\left(x_N^{(1)} + x_{N-1}^{(1)}\right)/2, & 1 \end{bmatrix}$$

Then

$$B=\begin{bmatrix}-204.875 & 1\\ -344.125 & 1\\ -475.625 & 1\\ -600.625 & 1\\ -723.625 & 1\\ -844.000 & 1\\ -969.125 & 1\\ -1108.375 & 1\end{bmatrix}$$

$y_N=[x_2^{(0)}, x_3^{(0)}, \ldots, x_N^{(0)}]^T$, Namely

$y_N=[141.25, 137.25, 125.75, 124.25, 121.75, 119.00, 131.25, 147.25]^T$

$\hat{a}=(B^TB)^{-1}B^Ty_N$

$\hat{a}=\begin{bmatrix}a\\u\end{bmatrix}=\begin{bmatrix}-0.0010\\131.6518\end{bmatrix}$

Then a =-0.0010, u = 131.6518. Substituting A and u to formula (2)

$$\hat{x}_{k+1}^{(1)}=\left(x_1^{(1)}-\frac{u}{a}\right)e^{-ak}+\frac{u}{a}$$

(2)

Formula (3) can be obtained.

$$x_{k+1}^{(1)}=126970.6628-126645.4128\,e^{0.0010k}$$

(3)

3.2. Prediction model of total number of accidents

The original discrete data sequence:

$x^{(0)}=\{307, 380, 296, 275, 228, 227, 283, 257, 233\}$

The statistical data are processed by moving average method.

$x^{(0)}=\{325.25, 340.75, 311.75, 268.5, 239.5, 241.25, 262.5, 257.5, 239\}$

The sequence can be obtained by a process of cumulative generation.

$x^{(1)}=\{325.25, 666, 977.75, 1246.25, 1485.75, 1727, 1989.5, 2247, 2486\}$

Data matrix B, yN can be obtained as follow.

$$B=\begin{bmatrix}-495.625 & 1\\ -821.875 & 1\\ -1112 & 1\\ -1366 & 1\\ -1606.375 & 1\\ -1858.25 & 1\\ -2118.25 & 1\\ -2366.5 & 1\end{bmatrix}$$

$y_N=[340.75, 311.75, 268.5, 239.5, 241.25, 262.5, 257.5, 239]^T$

Then $\hat{a}=\begin{bmatrix}a\\u\end{bmatrix}=\begin{bmatrix}0.0465\\338.4034\end{bmatrix}$

$$x_{k+1}^{(1)}=7272.9617-6947.7117e^{-0.0465k}$$

(4)

3.3. Optimization and improvement of prediction model

In order to improve the accuracy of the model, the parameters are fitted twice to improve the model [7, 8]. Taking the forecasting model of the number of accidents caused by illegal use as an example. The above time response equation is written as follows:

$x^{(1)}(k+1) = Ae^{0.0010k} + B$

"A" and "B" are estimated according to the value "a" of the first sequence estimation and the original sequence $x^{(1)}(k)$. The sequence matrix G and the sequence vector $x^{(1)}$:

$$G = \begin{bmatrix} e^0, & 1 \\ e^{-a}, & 1 \\ \cdots & \cdots \\ e^{-a(k-1)}, & 1 \end{bmatrix} = \begin{bmatrix} 1 & 1 \\ 0.9990 & 1 \\ 0.9979 & 1 \\ 0.9969 & 1 \\ 0.9959 & 1 \\ 0.9948 & 1 \\ 0.9938 & 1 \\ 0.9928 & 1 \\ 0.9917 & 1 \end{bmatrix}$$

$$x^{(1)} = [x_1^{(1)}, x_2^{(1)}, \cdots, x_N^{(1)}]^T,$$

namely

$x^{(1)} = [134.25, 275.50, 412.75, 538.50, 662.75, 784.5, 903.5, 1034.75, 1182.0]^T$

The parameters A and B are obtained.

$$\begin{bmatrix} A \\ B \end{bmatrix} = (G^T G)^{-1} G^T x^{(1)} = \begin{bmatrix} -124224.2923 \\ 124369.3079 \end{bmatrix}$$

The final response equation is obtained, which is shows in formula (5).

$$x^{(1)}(k+1) = -124224.2923e^{0.0010k} + 124369.3079 \quad (5)$$

$$x^{(0)}(k+1) = x^{(1)}(k+1) - x^{(1)}(k)$$

In the formula, $k = 0, 1, 2, 3, \ldots, N; x^{(0)}(1) = x^{(1)}(1)$. The calculation results are shown in Table 2.

Table 2. Calculating results of accidents caused by illegal use

Year	Serial number	Actual value	Actual first-order cumulative value	Predictive first-order cumulative value	Predictive value	Residual value
2008	1	134.25	134.25	145.016	145.016	-10.766
2009	2	141.25	275.50	273.753	128.737	12.513
2010	3	137.25	412.75	402.357	128.604	8.646
2011	4	125.75	538.50	530.828	128.471	-2.721
2012	5	124.25	662.75	659.166	128.338	-4.088
2013	6	121.75	784.50	787.370	128.205	-6.455
2014	7	119.00	903.50	915.442	128.072	-9.072
2015	8	131.25	1034.75	1043.381	127.939	3.311
2016	9	147.25	1182.00	1171.187	127.806	19.444
2017	10			1298.861	127.674	
2018	11			1426.403	127.542	
2019	12			1553.813	127.410	
2020	13			1681.090	127.277	
2021	14			1808.236	127.146	

2022	15			1935.250	127.014	

After the same calculation, the prediction model of the total number of accidents can be optimized. The optimization model is formula (6) and the calculation results are shown in Table 3.

$$x^{(1)}(k+1) = -6824.7803e^{-0.0465k} + 7167.9193 \quad (6)$$

Table 3. Calculating results of total number of accidents

Year	Serial number	Actual value	Actual first-order cumulative value	Predictive first-order cumulative value	Predictive value	Residual value
2008	1	325.25	325.25	343.139	343.139	-17.889
2009	2	340.75	666.00	653.415	310.276	30.474
2010	3	311.75	977.75	949.584	296.169	15.581
2011	4	268.50	1246.25	1232.289	282.705	-14.205
2012	5	239.50	1485.75	1502.141	269.852	-30.352
2013	6	241.25	1727.00	1759.725	257.584	-16.334
2014	7	262.50	1989.50	2005.598	245.873	16.627
2015	8	257.50	2247.00	2240.293	234.695	22.805
2016	9	239.00	2486.00	2464.318	224.025	14.975
2017	10			2678.158	213.840	
2018	11			2882.277	204.118	
2019	12			3077.115	194.839	
2020	13			3263.096	185.981	
2021	14			3440.621	177.525	
2022	15			3610.075	169.454	

4. Analysis of prediction results

Based on the above calculation and analysis, the total number of accidents and the number of accidents caused by illegal use of special equipment in 2008-2022 can be predicted. The total number of accidents and the number of accidents caused by illegal use of special equipment in 2008-2022 are estimated and predicted by the established prediction model.

The analysis results are shown in Fig. 1 and Fig. 2. It can be seen from the figure that the model fits well. From the Fig.2, it shows that the total number of accidents of special equipment is decreasing year by year, which is inseparable from the strict supervision system of special equipment in China. However, the number of accidents caused by illegal use does not change significantly, accounting for a large proportion, indicating that there are still large loopholes in the regulation of human abnormal behavior. In order to keep the accident situation of special equipment stable, Scientific and comprehensive quality supervision measures, formulate scientific safety production control indicators should be adopted, which are useful to ensure the safe operation of special equipment [9, 10].

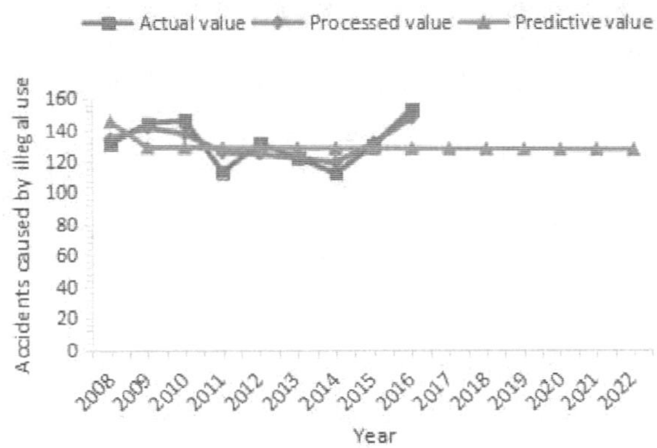

Fig.1 The values of accidents caused by the illegal use of special equipment

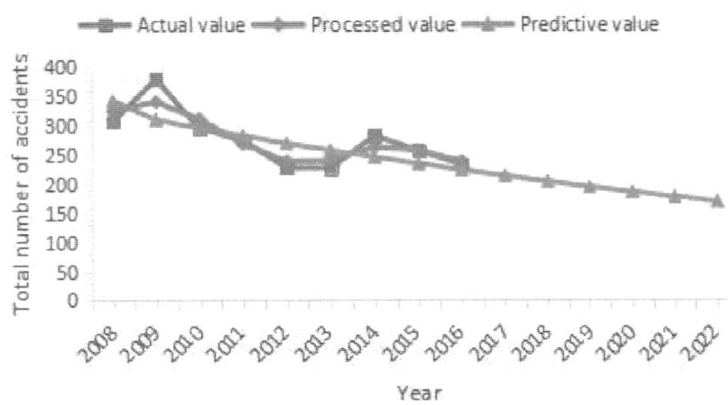

Fig.2 The values of the total number of accidents

5. Conclusions

In the theory and method of grey prediction, dealing with the original statistical data can eliminate the influence of random data to a certain extent. The grey model described by cumulative generation and differential equation can effectively deal with poor information and disordered data, and has good prediction accuracy and practicability in a certain period of time. This shows that the prediction method based on grey theory is suitable for the trend prediction of safety accidents of special equipment in China.

The main purpose of establishing the model is to provide scientific theoretical reference for relevant policies and strategic objectives of safety accident control. Grey theory needs relatively few raw data and has strong operability. Abnormal human behavior is the main factor in the occurrence of special equipment accidents, which needs to be monitored more strictly. In order to improve the safety status of special equipment and meet the needs of the development of special equipment, scientific safety supervision of special equipment is particularly important.

6. Reference

[1] Luo Hongqi, Nie Xuejun, Liu Xia. Mechanization profit portion estimation in agricultural industry based on the grey model[J]. *Journal of agricultural mechanization research*, 2011, (10):44-46

[2] Cui Qingling, Luo Yun, Cui Gang, et al. Study on preidiciton of special equipment accident based on grey theory[J]. *Journal of safety science and technology*, 2013, 9(5): 141-144

[3] Executive Editorial Committee of China Heavy Machinery Industry Yearbook. China Heavy Machinery Industry Yearbook [J]. Beijing: *China Machine Press*, 2009-2017

[4] Li Yuefen, Tang Jie, Li Yanmei. Evaluation of grassland soil quality by principal component analysis and grey correlation analysis [J]. *Global Geology*, 2004, 23 (2): 169-174

[5] Wu Xinmin, Pan Genxing. Relevance analysis of factors affecting heavy metal pollution in urban soils [J]. *Acta Pedologica Sinica*, 2003.40 (6): 921-928
[6] Ouyang Tao, Xie Xiaoyun. Statistics [M]. Changsha: *National Defense University of Science and Technology Press*. 2002
[7] Chen Lei. Application of gray theory in freeway traffic volume prediction [J]. *Liaoning communication science and technology*, 2004,(10): 37-39
[8] Lin Shaoxia, Lin Changhu, He Tengbing, et al. Application of grey relational degree method in analysis of factors affecting agricultural non-point source contaminated soil [J]. *Bulletin of Soil and Water Conservation*, 2010,30(4):171-174
[9] Kou Yi. Application of combination forecasting models in road freight turnover [J]. *China warter transport*, 2007,(11): 216-218
[10] Zhang Qinwen, Wang Xuemen. Grey System Analysis of Rural Economy - Model, Method and Application [M]. Beijing: *Academic Journal Press*. 1989

Acknowledgments
Project is supported by National Key R & D Program of China (2017YFF0207200).

Investigation on Polymer Composite Materials Wear Reinforced by Microparticles of Jatropha Curcas L. Waste

Miroslav Müller[1], Petr Valašek[1], Martin Tichý[1], Alessandro Ruggiero[2] and Roberto D´Amato[3]

[1]Czech University of Life Sciences Prague (CULS), Kamýcká 129, 165 00 Praha-Suchdol (Czech Republic)
[2]University of Salerno, Department of Industrial Engineering, Via Giovanni Paolo II, nº 132 84084 Fisciano (Italy)
[3]Departamento de Ingeniería Mecánica, Química y Diseño Industrial, E.T.S. de Ingeniería y Diseño Industrial – Tecnichal University of Madrid (UPM), C/ Ronda de Valencia, nº3 28012, Madrid (Spain)

Email: valasekp@tf.czu.cz

Abstract. The subject of the paper is a research on the wear resistance by friction against loosely fixed abrasive particles on a polymer composite material reinforced by waste microparticles from a pressing process of Jatropha curcas L. seeds with concentration from 5 to 20 wt.%. The seed pressing waste was used in form of Whole seeds cake (WSC), Seed shells (SS) and Seed kernels cake (SKC). The waste, which is difficult to further utilize for its inability to use as feed, arises at the pressing process of Jatropha Curcas L. seeds. A matrix of composite materials was epoxy-based. The wear tests were performed by the device Tester T-07 according to GOST 23.208-79. An addition of the filler significantly increased the wear resistance by friction against a rubber wheel with loosely fixed abrasive particles of sand. The wear resistance was increased up of 69% to 82% at tested polymer composite systems against matrix (resin). Composites materials with the filler Whole seeds cake (WSC) achieved the best results.

1. Introduction

The composite materials find applications in various field of industry production, where is necessary to consider not only a strength but also the wear resistance [1, 2, 3, 4, 5, 6, 7]. The wear resistance has own specification and understanding tribological behaviour of composite materials is important [8, 3]. For improvement the wear resistance various fillers based on micro and nanoparticles have been historically used [9].

A particle of SiC, Al_2O_3, ballotine, copper, aluminium and short carbon fibre, glass fibre etc. could be an example [4, 10, 11, 12, 13, 14]. An alternative filler based on natural materials, e.g. wood flour, microparticles from egg shell, nut shells, fruit stones etc. are used currently [15, 16, 17].

Use of the filler in a composite material field has three main reasons. The first one is a cost reduction, the second one is an increase of mechanical properties and use of the filler for improvement of the wear resistance. At the tribological behaviour of the composite material it is necessary to consider the wear type on a product surface and interactions between a matrix and reinforced phase inside of the composite structure [18, 5]. An inappropriate interaction between the matrix and the filler leads to increased wear of the material which is caused by a loose of the material, i.e. the mass loss [19].

The third one is the use of waste from agricultural commodities production, which is renewable source. One of a possibility is the material use of waste from Jatropha Curcas L. seed pressing, i.e. Jatropha Oil Cake. This waste proved its positives at tribological tests ball-on-flat with reciprocating motion with frequencies of 5 Hz and 10 Hz at polymer composite materials [20]. Main use of seeds from the plant Jatropha Curcas L. is oil production by pressing [21]. Oil cakes are secondary products after mechanical oil pressing which proved their strengths in practice [6]. The pressed oil and oil cakes are not eatable. For this reason, the secondary product is not suitable for farm animals feeding. At next treatment of Jatropha oil cakes, the filler arises which can be used in polymer composites. The example of the effective use can be seen e.g. in research by Shivamurthy [22] and Mohan [23], who proved the effectivity of the filler from oil cake in powder form e.g. improvement of tribological properties.

The aim of the research is an evaluation of the polymer composite wear resistance with the matrix based on the epoxide and the reinforced phase based on waste from Jatropha Curcas L. seed treatment, i.e. from oil pressing in form of Whole seeds cake (WSC), Seed shells (SS) and Seed kernels cake (SKC).

2. Materials and methods

The particle polymer composite with a continuous phase in a form of a two-component epoxy adhesive Lepox 1200 and a discontinuous phase (reinforcing particles) in a form of waste from pressing process of Jatropha Curcas L. seeds in a form of Whole seed cakes (WSC), Seed shells (SS) and Seed kernel cakes (SKC) was the subject of the experiment. A distribution of the input raw material for a preparation of the filler came out from the secondary product at oil production.

WSC, SS and SKC were desiccated at the temperature 105 ± 5°C for 24 hours and then powdered by multi-stage crushing machine. This input material was sorted by the sieve analysis. For research a smallest fraction was used which fell into sieve with size 100 μm of mesh. The size of particle fillers WSC, SS and SKC were measured by means of an image analysis with programme Gwyddion and images from electron microscope (SEM). The size of WSC was 72.28 ± 38.5 μm, size of SS was 62.42 ± 28.6 μm and size of SKC was 68.56 ± 42.8 μm.

Test samples for the wear resistance evaluation according to standard GOST 23.208-79 (Ensuring of wear resistance of products. Wear resistance. Testing of materials by friction against loosely fixed abrasive particles) were made in the form from silicone rubber Lukopren N 1522. The test samples had dimension 25 x 25 x 5 mm. The composite was made from the epoxide and the reinforcing phase with concentration 5, 10, 15 a 20 wt%. The silicone form was filled by composite mixture and cured for 24 hours at laboratory temperature.

The evaluation methods of composite materials with the filler WSC, SS and SKC were based on self-comparison of the wear resistance according to standard GOST 23.208-79. The test of the wear resistance was measured according to standard GOST 23.208-79 i.e. on device with rubber wheel (fig. 2 A). The worn test samples with the characteristic wear track are shown in fig. 2 B, C, D. The float sand Otava with particles of the size 0.2 to 0.315 mm fell between the test sample and rubber wheel. The test samples were load by 680 g. The test samples were cleaned with compressed air before and after the test. The mass loss was measured by weighing-machine KERN ABS with tolerance 0.1 mg. The test samples were measured at 100 turns of the rubber wheel. For each measuring three samples were used. The test samples after wear test are shown in fig. 2 B, C and D, where a typical relief of the worn surface is noticeable.

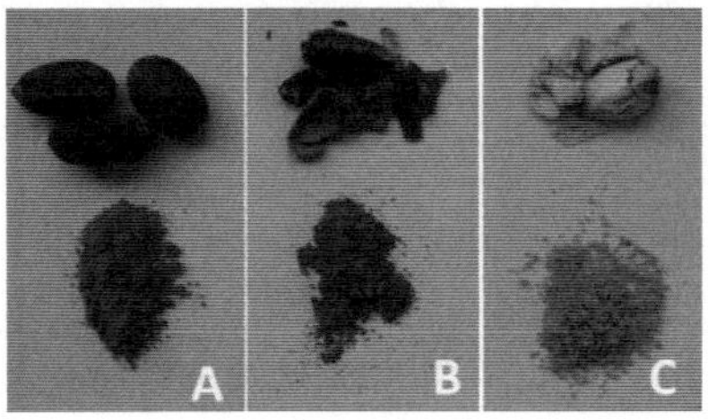

Figure 1. Filler obtained as waste from pressing of Jatropha Curcas L. seeds for oil production: A: Whole seed cakes (WSC), B: Seed shells (SS), C: Seed kernel cakes (SKC)

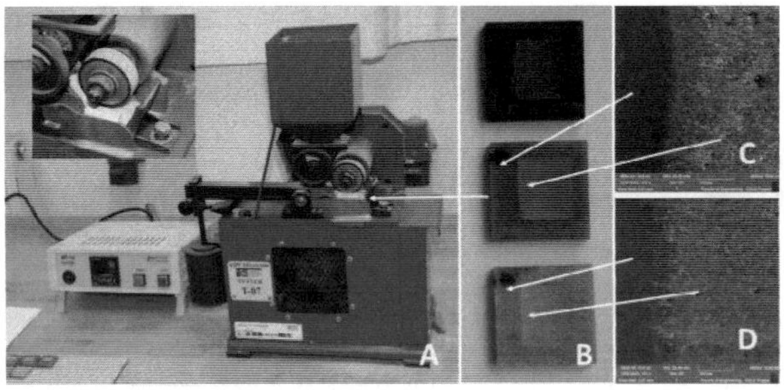

Figure 2. Measuring of wear resistance by friction against loosely fixed abrasive particles according to GOST 23.208-79: A: Device Tester T-07, ITC PIB Institute of Technology Radom and control device BT – 16 Controller, B: test samples after wear resistance test, C: SEM image of test samples with filler Whole seed cakes (WSC), MAG 120 x, D: SEM image of test samples with filler Seed kernel cakes (SKC), MAG 122 x

3.Results and discussion

By the electron microscopy the shape, the size and the texture surface of the fillers Whole seeds cake (WSC), Seed shells (SS) and Seed kernels cake (SKC) were observed, see fig. 3. The fillers WSC, SS and SKC are different among themselves in the shape characteristic, especially Seed kernel cakes have very rugged surface composed of many segments.

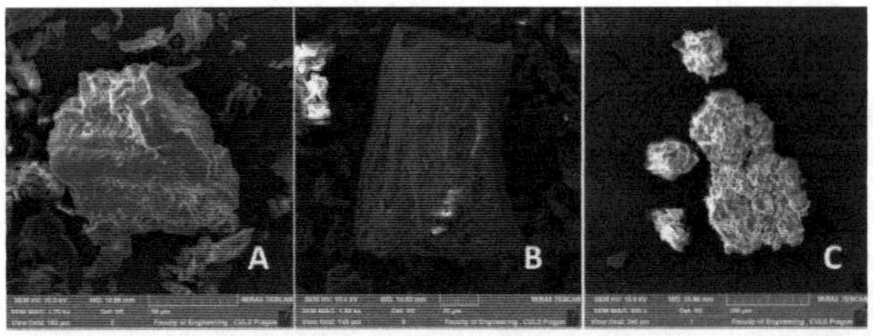

Figure 3. SEM images of filler obtained as waste from pressing of Jatropha Curcas L. seeds for oil production: A: Whole seed cakes (WSC), MAG 1.70 kx, B: Seed shells (SS), MAG 1.86 kx, C: Seed kernel cakes (SKC), MAG 800 x

The process of the matrix wear (resin, without the filler) depending on the wear track is shown in fig. 4. The comparison of composite material results in fig. 5, 6 and 7 points on the significant influence for the wear resistance improvement of polymer composites. The average value of the worn matrix (resin) is 51.99 ± 6.95 mg. The composite materials reported much smaller values of the wear, i.e. the mass loss was in interval from 9.49 to 15.81 mg. The results comparison is evident from fig. 8.

The process of the wear depending on the track is shown in fig. 5, 6 and 7 which present a cumulative mass loss depending on a trajectory. The results point on a linear increase of the mass loss of tested samples. For the relevant evaluation it is necessary to determine the strength (intensity) of the linear dependence which is the aim of correlation analyses.

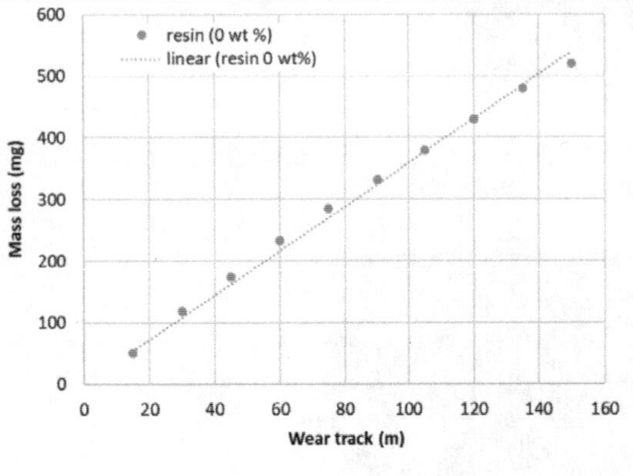

Figure 4. Matrix wear process (resin)

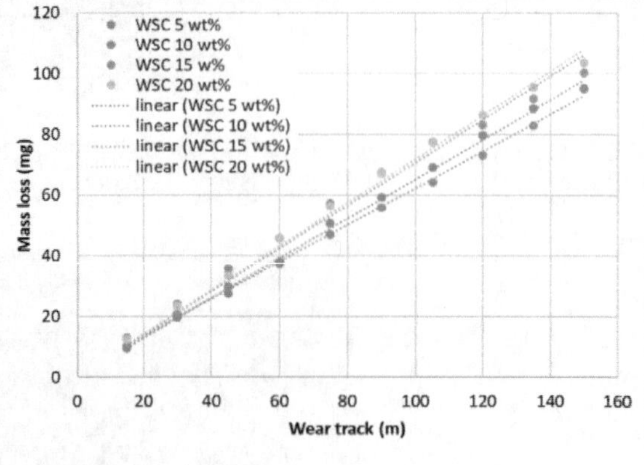

Figure 5. Wear process of composite material with filler Whole seed cakes (WSC)

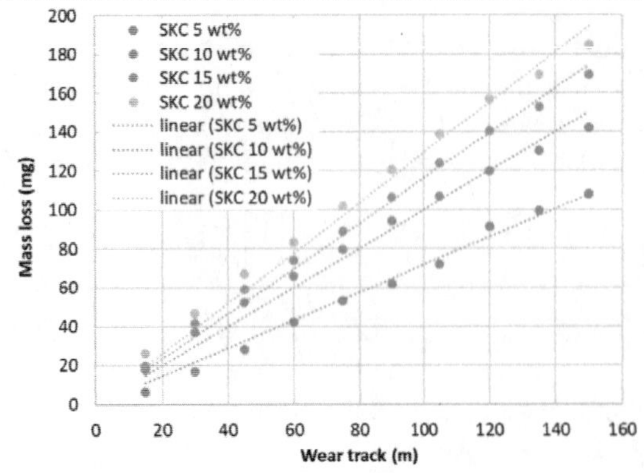

Figure 6. Wear process of composite material with filler Seed kernel cakes (SKC)

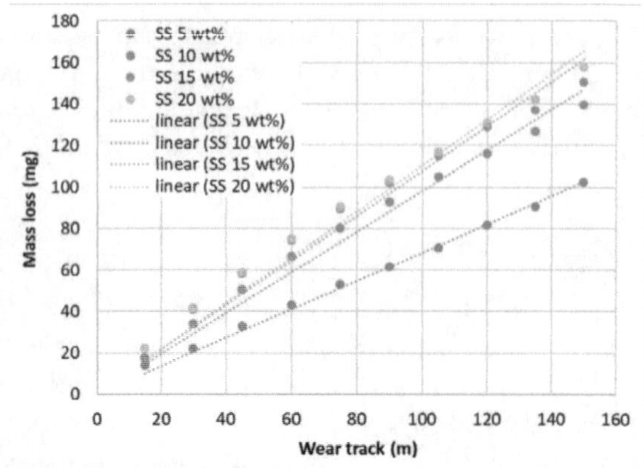

Figure 7. Wear process of composite material with filler Seed shells (SS)

Each filler concentrations of composite materials were tested among themselves according to ANOVA F-test in significance level 0.05. The results of the statistical comparison are shown in fig. 8. Composite materials showed less dispersion than the matrix. For null hypothesis H_0 the status was marked when there is no statistically significant difference among the data sets being compared in terms of their mean values: $p > 0.05$. The hypothesis H_0 was not confirmed at composite materials with filler Seed shells (SS) and Seed kernel cakes (SKC), i.e. there is the significant difference ($p = 0.000$) among the tested materials depending on the different filler concentrations in the significance level 0.05. The hypothesis was confirmed on composite with filler Whole seed cakes (WSC), i.e. there is no difference on the wear resistance among tested concentrations ($p = 0.098$).

The best and at the same time balanced results were achieved on the composite material with the filler Whole seed cakes (WSC) which contained hard particles of Seed shells (SS) and soft particles of Seed kernel cakes (SKC) which contained increased percentage of oil.

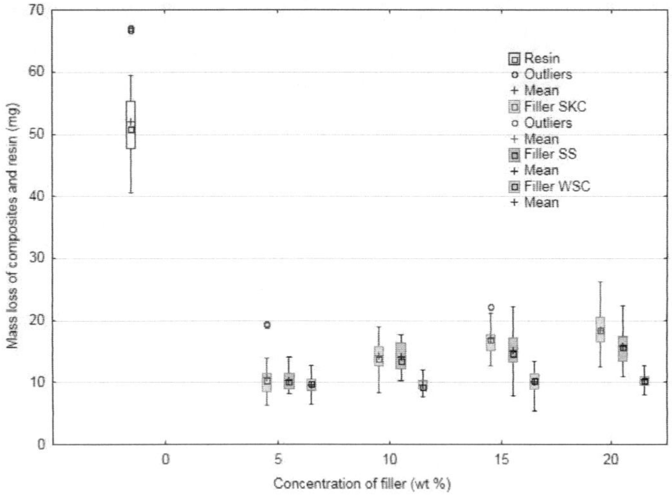

Figure 8. Statistical comparison of measured results of composite materials

The surface of the worn composite material with the filler Whole seed cakes (WSC) is shown in fig. 9. In fig. 9 A the texture of the worn surface at MAG 5.00 kx is noticeable. In fig. 9 B (MAG 1.60 kx) the detail on a delamination of larger particle from the matrix is noticeable. The delamination of larger particles decreases the wear resistance. In fig. 9 C larger particle with good adhesion force is visible which did not allow it to release from the matrix. Whole seed cakes (WSC) contain soft particles of a core and hard particles of a peel.

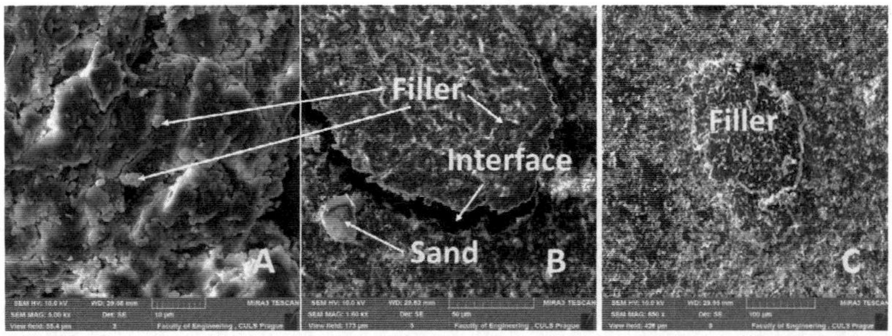

Figure 9. SEM images of worn surface according to GOST 23.208-79 of composite material with Whole seed cakes (WSC)

The worn surface of the composite material with the filler Seed shells (SS) is shown in fig. 10. Seed shells (SS) are hardener than Seed kernel cakes (SKC). On the worn surface a significant groove was examined - see fig. 10 A (MAG 500 x) with detail in fig. 10 B (MAG 500 x) and C (MAG 3.26 kx).

Figure 10. SEM images of worn surface according to GOST 23.208-79 on composite material with filler Seed shells (SS)

The worn surface of the composite material with the filler Seed kernel cakes (SKC) is noticeable on fig. 11. This filler contains the largest percentage of residual oil from pressing process and contains smaller particles, see fig. 3 C. In fig. 11 A (MAG 1.50 kx) a typical relief of the worn surface is obvious. At the filler Seed kernel cakes (SKC) larger filler cluster was pulling out, see fig. 11 B (MAG 5.00 kx), which is subsequently applied to the surface in the direction of a disc rotation due to the pressure force of the rubber disc during the wear process, see fig. 11 C (MAG 100 x).

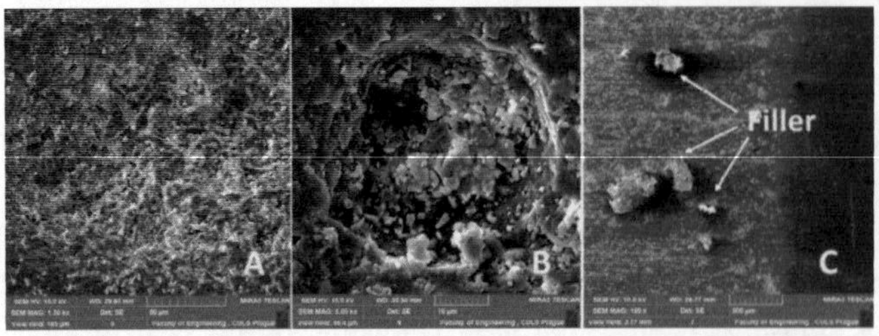

Figure 11. SEM images of worn surface according to GOST 23.208-79 on composite material with filler Seed kernel cakes (SKC)

The research results confirmed the fact that the filler addition into the thermosetting matrix based on epoxide influenced the mechanical properties of the composite [13, 14, 24, 25]. The significant factor is the type and the dimension of the filler [10]. The conclusions of many authors regarding the possibility of effective use of the filler based on waste from agricultural products processing in the area of polymer composite materials were certified [6, 15, 16, 17, 19].

4. Conclusions

The polymer composite materials based on waste from pressing process of Jatropha Curcas L. seeds in the form of Whole seed cakes (WSC), Seed shells (SS) and Seed kernel cakes (SKC) with concentration from 5 to 20 wt.% proved the significant wear resistance by friction against the rubber wheel with loosely fixed abrasive particles of sand according to GOST 23.208-79.

The wear resistance of polymer composite test samples was increased to the matrix (resin) up of 69% to 82%. The best results reported the composite material with the filler Whole seed cakes (WSC). The statistical analysis of this filler proved that the wear resistance was not directly dependent on the filler concentration in interval 5 to 20 wt.%. This filler reported also the least dispersion of results. These results are probably caused by microparticles combination of soft filler Seed kernel cakes (SKC) and hard filler Seed shells (SS).

Acknowledgement

Supported by Internal grant agency of Faculty of Engineering, Czech University of Life Sciences Prague, No.2019:31140/1312/3108

References

[1] N.S.M. El-Tayeb, B.F. Yousif, T.C. Yap. Tribological studies of polyester reinforced with CSM 450-R-glass fiber sliding against smooth stainless steel counter face. *Wear* **261**, 443–452 (2006)

[2] B.N. Ravi Kumar, B. Suresha, M. Venkataramareddy. Effect of particulate fillers on mechanical and abrasive wear behaviour of polyamide 66/polypropylene nanocomposites. *Mat. and Desig.* **30**, 3852–3858 (2009)

[3] P. Valášek, M. Müller. Abrasive wear in three-phase waste-based polymeric particle composites. *Tehnicki Vjesnik-Technical Gazette* **12** (2), 257-262 (2015)

[4] M. Müller, R. Chotěborský, P. Valášek, S. Hloch. Unusual possibility of wear resistance increase research in the sphere of soil cultivation. *Tehnicki Vjesnik-Technical Gazette* **20** (4), 641-646 (2013)

[5] A. Ruggiero, M. Merola, P. Carlone, V.M. Archodoulaki. Tribo-mechanical characterization of reinforced epoxy resin under dry and lubricated contact conditions. *Compos. Part B: Eng.* **79,** 595–603 (2015)

[6] A. Ruggiero, R. D'amato, M. Merola, P. Valášek, M. Müller. Tribological characterization of vegetal lubricants: Comparative experimental investigation on Jatropha curcas L. oil, Rapeseed Methyl Ester oil, Hydrotreated Rapeseed oil. *Tribology. Int.* **109**, 529-540 (2017)

[7] Ch.G. Papakonstantinou, D. Bekiaris. Comparison of Shear Design Models for Reinforced Concrete Beams Strengthened with Inorganic Matrix Composite Materials. *International Journal of Materials, Mechanics and Manuf.* **6** (1), (2018)

[8] K. Friedrich, P. Reinicke. Friction and wear of polymer based composites. *Mech. Compos. Mater.* **34**, 503–514 (1998)

[9] B. Suresha, B.N. Ravi Kumar. Two-body Abrasive Wear Behavior of Particulate Filled Polyamide66/Polypropylene Nanocomposites. *J. of Appl. Polymer Sc.* **119**, 2292-2301 (2010)

[10] F. Shao-Yun, F. Xi-Qiao, L. Bernd, M. Yiu-Wing. Effects of particle size, particle/matrix interface adhesion and particle loading on mechanical properties of particulate–polymer composites. *Compos. Part B* **39,** 933–961 (2008)

[11] S. Basavarajappa, A.G. Joshi, K. Arun, A.P. Kumar, M.P. Kumar. Three-Body Abrasive Wear Behaviour of Polymer Matrix Composites Filled with SiC Particles. *Polymer-Plastics Techn and Eng* **14**, 8-12 (2010)

[12] H. Pihtili, N. Tosun. Effect of load and speed on the wear behaviour of woven glass fabrics and aramid fibre-reinforced composites. *Wear* **252**, 979–984 (2002)

[13] M. Müller, P. Valášek. Abrasive wear effect on Polyethylene, Polyamide 6 and polymeric particle composites. *Manufact Tech* **12**, 55-59 (2012)

[14] D. Galusek, J. Sedláček, R. Riedel. Al2O3-SiC composites prepared by warm pressing and sintering of an organosilicon polymer-coated alumina powder. *J of the Europ. Ceramic Society* **27** (6), 2385-2392 (2007)

[15] M. Zieleniewska, K.M. Leszczynski, L. Szczepkowski, A. Bryskiewicz, M. Krzyzowska, K. Bien, J. Ryszkowska. Development and applicational evaluation of the rigid polyurethane foam composites with egg shell waste. *Polymer Degradation and Stability* **132**, 78-86 (2016)

[16] H.Y. Cheung, M.P. Ho, K.T. Lau, F. Cardona, D. Hui. Natural fibre-reinforced composites for bioengineering and environmental engineering applications. *Compos. Part B.* **40**, 655-663 (2009)

[17] V. Matějka, Z. Fu, J. Kukutschova, Q. Shicheng, S. Jiang, X. Zhang, R. Yun, M. Vaculik, M. Heliova, Y. Lu. Jute fibers and powderized hazelnut shells as natural fillers in non-asbestos organic non-metallic friction composites. *Mater. and Desig.* **51**, 847–853 (2013)

[18] N.S.M. El-Tayeb, B.F. Yousif. Evaluation of glass fibre reinforced polyester composite for multi-pass abrasive wear applications. *Wear* **262**, 1140–1151 (2007)

[19] M.N. Satheesh Kumar, Z. Yaakob, N. Mohan, S.P. KumareshBabu, S. Siddaramaiah. Mechanical and Abrasive Wear Studies on Biobased Jatropha Oil Cake Incorporated Glass–Epoxy Composites. *J. of Appl. Polymer Sc.* **124**, 484–494 (2012)

[20] A. Ruggiero, P. Valášek, M. Müller, R. D´Amato. Tribological investigation of epoxy/seed particle composite obtained from residues of processing Jatropha Curcas L. fruits. *Composites Part B: Engineering*. **167**, 654-667 (2019)

[21] D. Herák, J. Blahovec, A. Kabutey. Analysis of the axial pressing of bulk Jatropha curcas L. seeds using reciprocal slope transformation. *Biosystems Eng.* **121**, 67-76 (2014)

[22] B.A.B. Shivamurthy, K.A. Murthy, P.C.A. Joseph, K.A. Rishi, K.U.B. Bhat, S.B. Anandhan, Mechanical properties and sliding wear behavior of jatropha seed cake waste/epoxy composites. *J. of Mater. Cycles and Waste Management* **17** (1), 144-156 (2004)

[23] N. Mohan, S. Natarajan, S.P. KumareshBabu Siddaramaiah. Investigation on sliding wear behaviour and mechanical properties of jatropha oil cake-filled glass-epoxy composites. *J. of the American Oil Chemists' Society* **88**, 111–117 (2010)

[24] H.S. Kim, M.A. Khamis. Fracture and impact behaviours of hollow micro-sphere/epoxy resin composites. *Composites Part A: Appl. Sc. and Manuf.* **32** (9), 1311-1317 (2001)

[25] B. Agoudjil, L. Ibos, J.C. Majeste, Y. Candau, Y.E.P. Mamunya. Correlation between transport properties of Ethylene Vinyl Acetate/glass, silver-coated glass spheres composites. *Compos. Part A: Appl. Sc. and Manuf.* **39** (2), 342-351 (2008)

IOP Conf. Series: Materials Science and Engineering **638** (2019) 012012 doi:10.1088/1757-899X/638/1/012012

Numerical Investigation of Hybrid of Eglass and Basalt Fiber Reinforced Epoxy Tube Pressurized Internally

Thamir A. D. M. S. Almula[1*], Ahmad Hassan Khuder[2], Mohd Yazid Yahya[3] and Amran Ayob[3]

[1]Engineering Technical College/ Mosul, Northern Technical University NTU, 41002 Mosul, Iraq
[2]Electromechanical Engineering department /University of Technology, 10066 Baghdad, Iraq
[3]Center for Composites, Universiti Teknologi Malaysia, 81310 Skudai, Johor, Malaysia

*Email: thamir_own@yahoo.com; thamir_own@ntu.edu.iq

Abstract. The composite hybrid pipe is a novel structure made of concentric composite multilayer possess improved thermal and mechanical properties better than its original components. At variance experimental analysis, the numerical studies on the structure and behavior of composite hybrid natural fiber reinforced epoxy pipe materials pressurized internally seem lacking. In this study, numerical analysis was carried out for three different stacking plies of composite hybrid natural fiber ±55° basalt and Eglass fiber reinforced epoxy tube which was tested under the subjection for three modes of applied load, the hoop and longitudinal tensile tested as well as subjection under biaxial internal pressures load in accordance to ASTM's standard to investigate the mechanical behavior and the optimal configuration among them. Also, analyzed the comparable three types of hybrid natural fiber (basalt)/ Eglass reinforced epoxy with pure ±55° Eglass reinforced epoxy and pure ±55° basalt reinforced epoxy numerically to predict the performance for the hybrid over the pure E-glass and basalt fiber reinforced epoxy composite pipe as a good replacement to meet the different demand of applications in the pressure pipelines. Also The three stacking plies types of hybrid composite pipes were compared with the pure E-glass and basalt fiber reinforced epoxy pipes. All samples were fabricated using a dry filament wound CNC machine with ±55° orientation angle, then infused with epoxy resin using vacuum infusion procedure (VIP). Numerical analysis was carried out using finite element commercial code ANSYS V14.

1. Introduction

Composite hybrid pressure pipes are advanced structures produced by concentric layers of wound fiber reinforced matrix possess better thermal and mechanical properties exceeds of their constituent materials. Large demands for composite pressure pipe products have been motivated by the satisfying for the requirements of erosion and corrosion, high stiffness properties with lighter weight and low price, especially for the composite natural fiber used for its abundance, cheap and healthy with environmental. Composite hybrid pipe investigation studies had been increasingly popular and attracted many researchers because of its important applications in many fields, especially cooling water pipeline, aerospace, oxygen pipeline usage in hospitals, manufacturing power plants, biomaterials and other applications. Various sectors can utilize basalt like marine [1], ballistic warfare and impact [2,3]. The beneficial properties of basalt include its tensile strength and modulus. Basalt has proved itself as a viable option since it is a natural product with minimum environmental contamination, has light weight and fits well within the costs of the processing and hence can be used in different ways [4].

Usually, the process that yields a hybrid novel composition involves combining two foreign materials which are placed within a host matrix. Since, the combination of two materials produces an end product which has better properties like ductility, elasticity, resistance against fire and decreased weight [5-7]. New combinations can be developed by this method. Polyethylene (PE) fibers were incorporated with carbon fibers in the framework of an epoxy matrix by Park and Jang [8] and a hybrid laminated composite material was produced. The scientists decided to use PE fibers for this study since it has good stiffness and is a strong fiber that can withstand high degrees of stretching. This experiment showed that the position at which the reinforcing fiber is placed is the determining factor of the mechanical strength possessed by the compound produced by the combination process. The developed compound had a higher degree of strength when the position of CF was at the outermost layer. These findings reinforced the need to develop fibers that are lightweight yet strong, but are also long lasting and do not increase the cost of the production process. Basalt fibers are used most commonly in the present scenario since they are inorganic in nature, have good mechanical strength, can withstand high temperature and function in the presence of chemicals, retain their stability in different environments, require simple processing steps and cause less environmental pollution. The flowing lava which is produced from basalt based molten volcano rock is a rich source of basalt fibers [1,9]. Basalt stands out in comparison to E glass in terms of tensile strength and ability to withstand compression and costs quite less than fibers of carbon origin [10-12].

Replacing E-glass fiber by natural fibers as suitable substitutes can be recycled and comparable like flax, hemp, kenaf and sisal as well as renewability, abundant and cheap with contributes their high toughness, easy chapping, low density, capable of neutrality with CO2 and simplest fabrication procedure and maintenance [13]. Moreover the important request for reducing the need of petroleum derivatives products by nature and friendly replacement products is an interesting for many researchers and in developing. Subsequently, the investigations for the new products of natural fiber mechanical behavior have a great deal of attention for many researchers recently [14]. Lately, a step forward had been developed polymer composite material by natural fiber usage to avoid human dangerous from glass fiber [15].

So, basalt fibers are being explored as a viable option to act as a reinforcement compound in the development of hybrid composites or laminates [16]. There are many instances in the literature that show the use of basalt fibers as reinforcements in CFRP based composites. Lopresto et al.[17] compared plastic polymers reinforced with basalt fibers (BFRP) and plastics reinforced with glass fibers (GFRP) on the basis of Young's modulus, compressive strength and flexural behavior and their findings favored the use of basalt over glass. Manikandan et al. [18] also conducted a study which provided similar conclusion. Such a combination would not only be financially attractive, but would also be long lasting and weigh less, thus increasing the scope of basalt in the hybrid composite sector [19]. According to Carmisciano et al. [10] composites reinforced by basalt fibers [BWFRC] had higher shear strength and flexural modulus. The construction sector has made use of basalt fibers in the recent past and it has also been utilized in concrete materials as an external or internal reinforcement [20, 21, and 3]. Both tensile and flexural strength showed an increase of 16% when laminates made of basalt or epoxy TM were used, the tensile modulus would have an increment of 27% and flexural modulus would have an increment of 153% [22].

For easy and fast biaxial strength predicting method of the filament winding E-glass/epoxy tube is by using "netting analysis". This process was defined by a study of Hull and was extensively used throughout the primary progression of filament wound method [23]. It has been getting to give an equally rupture strength respectable approaches of composite tubes. This conventional approach and its simplicity may be the motives for its current use. Figure.1 presents the principal stresses are (σ_1 and σ_2), whereas the reference stresses are (σ_y, σ_x and τ_{xy}). $\sigma_y = \sigma_2 \sin^2 \theta, \sigma_x = \sigma_1 \cos^2 \theta$;

$\tau_{xy} = \sigma_1 (\sin\theta * \cos\theta)$; if $\sigma_H / \sigma_A = 2$; and $\sigma_y = \sigma_H$, $\sigma_x = \sigma_A$; then, within pipes case of closed ended with thin walls, $\sigma_H / \sigma_A = (\sigma_1 \sin^2\theta) / (\sigma_1 \cos^2\theta)$, then it will equal to ($\tan^2\theta = 2$) to give $\theta = \sqrt{2} = 54.7^o$

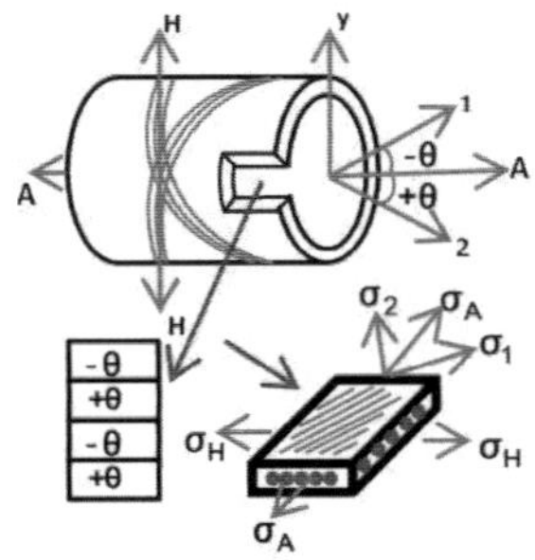

Figure 1. composite pipe Configuration {reference and principal axes (x, y) (1, 2) respectively}

The netting analyses can be used to predict the best angle that agrees with axial to hoop stress relative ratio. The best ratio of the glass/epoxy pipes is at twice of a hoop stresses to longitudinal stresses ratio, wound with ±55° orientation angle.

Xia et al [24] used an exact solution to study the different strain and stress in a tube that has many layers and has been pressurized internally; the basic presumption behind the model is that the tube has orthotropic for each layer. Later Xia et al [25] also used an exact technique in context of the sandwich tube with thick walls with internal pressure and thermo mechanical load and calculated the thermal stresses in such a system. Rosenow [26] developed an approach for calculating stresses and strain in case of composite cylinders composed of different layers in different angles varying from 15 degrees to 85 degrees. Accordingly, for thin layered composite tubes 55 degrees would be the best angle of orientation on the basis of comparison of actual and experimental data, but Xia et al [25] proposed 75 degrees as the best angle in case of internal pressure loading without axial.

The work done by Shultz's et al [27] showed that as far as thin pressure vessels that are manufactured from carbon fiber or epoxy are concerned the best winding angle was ±50 degrees. Highton et al [28] conducted research on failure and concluded that in case of epoxy tubes reinforced with glass ±75 degrees winding angle was sufficient to cause fracture under biaxial loads. Echold et al [29] has proposed the theory that it is not merely the anisotropic factors that dictate the chances of failure; it is also the type of raw materials used, the process of manufacturing and the testing conditions that play a very important role in deciding the failure limits. Carswell [30] has concluded that the leak pressure is directly proportional to resin flexibility. Spencer et al. [31] showed that when biaxial load was used to create a pressure and the axial negative strain that was used came out to be fewer than ±35 degrees, then the maximum value of leak pressures was detected to be ±55 degrees. The calculation of stresses and strain in the filament winding tube which were pressurized biaxialy internal pressures and could work in different temperature was facilitated by the work of Bakaiyan et al [32]. Due to this work the structure of the pipe was pictured as an orthotropic cylinder form in 3 dimensions with reinforced layers with alternate pigment composition. The conventional designs are based on standard safety mechanisms that are based on non-predictable problems during manufacturing, loading or problems arising out of mechanical or material issues.

The work of Wang et al. [33] focused on developing an enhancement method for BFRP using hybridization and the design was structured in a manner that it was best suited for use in long span cable based bridge. According to the findings of the study, there was an increase in the mechanical strength, modulus and fatigue behavior, the increase was attributed to hybridization. Zhang et al [34] also conducted a similar study. Quartz content up to 20% of basalt make it with excellent mechanical properties [35]. With augmented requests for fiber hybridization of composite pipe, the mechanical behavior sound investigation knowledge subjected to internal pressure is being necessary; consequently this numerically analysis would inspect the mechanical performance for hybrid (basalt and Eglass) filament winding multilayer fiber ±55° reinforced epoxy subjected to internal pressure, longitudinal and hoop tensile test load to verify the best hybrid structure among others. Also, compared with the pure wound ±55° Eglass and pure basalt fiber reinforced epoxy composite pipe to show the performance of hybridization over the conventional Eglass fiber reinforced epoxy and basalt fiber reinforced epoxy pipe.

2. Materials

In this numerical study the pipe profile and geometry used in the analysis can be shown in Figure 2. The five used materials were used, basalt/E-glass/basalt reinforced epoxy, E-glass/basalt/E-glass reinforced epoxy, hybrid (both of one E-glass strand and one basalt strand were wound altogether) reinforced epoxy, pure E-glass/epoxy and basalt/epoxy. All mentioned sample types were wound with the filament wound machine with three layers and 55° orientation angle of 2.5mm pipe thickness and infused with epoxy resin by the vacuum infusion process. Each sample dimensions followed the ASTM for each test that were subjected to. Three types of applied load were subjected to all mentioned sample types. The hoop and longitudinal tensile test and biaxial internal pressure test according to ASTM D2290[36], ASTM D2105[37] and ASTM D1599[38] respectively. The tubes were produced for experimentally tests, even though the investigation is numerical.

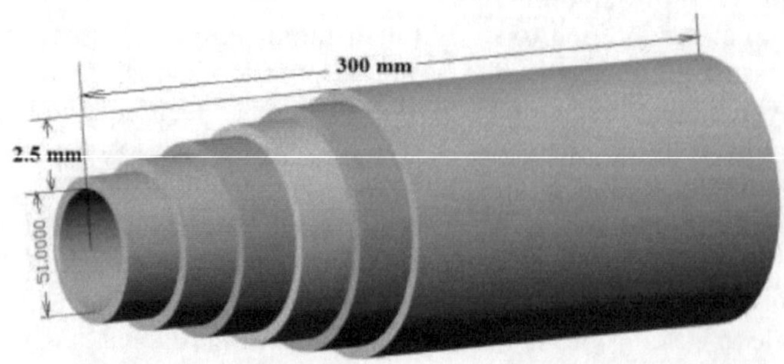

Figure 2. Dimensions and geometry of the composite tube.

According to ASTM D2584 [39] The volumetric fraction has been computed are by adopting ignition-loss of the cured strengthened resins, Following Jones [40-41]. The required parameters computed are by adopting the material mixture law (Eqations.1). Table 1 presented the two fiber type's properties which were used for the different composite constituent fabrication.

Table 1. Fiber characteristics for components of composites structures

	E [GPa]	G [GPa]	ρ (kg/cm³)	υ
Basalt	89.0	21.70	2.750	0.280
E-glass	72.4	29.70	2.540	0.330
Epoxy	3.60	1.440	1.20	0.280

Elements that were essential to simulate the hybrid tubes were: volume *v* in cm3, mass *M* in gm, volume fraction *V* and density ρ in g/cm³. The tube material mechanical properties that are computed are by adopting Eqns. 2-8 are: modulus of elasticity E, the density ρ, the Poisson ratio υ and shear moduli *G*. Table 2 displays the composite tube mechanical properties.

$$M_c = M_m + M_f \; ; v_m = M_m / \rho_m \; ; v_f = M_f / \rho_f \; ; v_c = v_m + v_f \; ; V_m = v_m / v_c \; ; V_f = v_f / v_c \quad (1)$$

$$\rho_c = V_f \cdot \rho_f + V_m \cdot \rho_m \quad (2)$$

$$E_{11} = E_f . V_f + E_m . V_m \quad (3)$$

$$\upsilon_{12} = \upsilon_{13} = \upsilon_f . V_f + \upsilon_m . V_m \quad (4)$$

$$E_{12} = (\xi\,\eta_c\,V_f + 1)/(1 - \eta_c\,V_f)\ ;\ \text{where}\ \eta_c = \frac{(E_f/E_m) - 1}{(E_f/E_m) + \xi};\ \xi = 2\ \text{for}\ E_{12}\ \text{computing} \tag{5}$$

$$G_{12} = \frac{1 + \xi\,\eta_g\,V_f}{1 - \eta_g\,V_f}\ ;\quad \text{where}\quad \eta_g = \frac{(G_f/G_m) - 1}{(G_f/G_m) + \xi};\ \xi = 1\ \text{for}\ G_{12}\ \text{computing} \tag{6}$$

$$G_f = E_f / 2(1 + \upsilon_f)\ ;\quad \text{and}\quad G_m = E_m / 2(1 + \upsilon_m)\ ;\quad \text{assume}\ G_{12} = G_{13}. \tag{7}$$

$$\upsilon_{23} = \upsilon_f\ .\upsilon_m\ / V_m\ .\upsilon_f + V_f\ .\upsilon_m \tag{8}$$

Table 2. Composite constituent material's characteristics

Composites	ρ [kg/m^3]	E_1 [GPa]	E_2 [GPa]	E_3 [GPa]	G_{12} [GPa]	G_{23} [GPa]	G_{13} [GPa]	ν_{12}	ν_{23}	ν_{13}
Eglass/Epo	2.54	45.8	10.1	10.1	3.46	3.46	3.46	0.308	0.31	0.308
Basalt/Epo	2.75	54.4	9.8	9.8	3.24	3.24	3.24	0.28	0.28	0.28
Hyb/Epoxy	2.645	50.2	9.99	9.99	3.35	3.35	3.35	0.29	0.29	0.29

3. A finite element simulation

Through utilizing a FEM-ANSYS V14 commercial software code, linearly and elastic composite properties investigation of the composite structure of the filament winding tube pressurized internally, axial tensile and hoop tensile test were implemented numerically. The hybrid of Eglass and Basalt fiber reinforced epoxy composite tube had been taken under consideration to be orthotropic. The fiber elastic property and ply angle were familiarized. The model with fine-meshed, linear 3D tetrahedral of thin wall thickness with 8 nodes were utilized for simulation for all load modes through divers of element numbers and node numbers for each load mode. For the internal pressure test mode, the element numbers were 416 and the nodes were 818, although for the hoop tensile testing, the elements were 4059 and the nodes were 16927, then for axial tensile testing the element numbers were 454 and the nodes were 465. The numerical analysis provided axial and hoop strain and stress for diverse fiber ply angle and for the three different types of hybrid composite selected material pipes. From the outcomes it can be assessed the behavior of the three different types of the novel hybrid basalt/E glass fiber reinforced epoxy to designate the optimal ply sequence among the three different types.

4. Results and discussions

The elastic and linear numerical investigations had been simulated utilizing three different modes of subjection load of biaxially pressurizing internally as well as longitudinal tensile and hoop tensile tests on three different kinds of composite hybrid stacking plies tubes of basalt/Eglass/basalt, Eglass/basalt/Eglass and hybrid contributed fibers of Eglass and basalt at the same time for the dry filament winding process assuming for all selected stacking hybrid plies of fiber materials wound with an angle of ±55° ply angles and six stacked layers. Figure.3 (a) displays the strain results contour of composite ±55° hybrid basalt/Eglass/basalt subjected to 18 MPa biaxial internally pressure. Hoop strain results contour can be shown in Figure.3 (b) for the subjection under hoop tensile load mode. According to ASTM D2105 standard that had been adopted so that there were no bend happened on the dogbone segment. Axial strain result contour shown in Figure.3 (c) under the subjection of axial tensile load mode.

Table 3. Shows the numerical results of longitudinal and hoop strains and stress against the diverse comparable three types of hybrids of basalt/Eglass/basalt, Eglass/basalt/Eglass and hybrid contributed fibers of Eglass as well as pure basalt and pure Eglass fiber reinforced epoxy tubes (a) under 18 MPa biaxial internally pressure, (b) under hoop tensile test and (c) under Longitudinal tensile test. The strain results of basalt/Eglass/basalt reinforced epoxy hybrid pipe type shows low differences in stress as much as strain under same internal pressure level. As well as the differences between the three different composite hybrid kinds for the simulated tube structures are because of their different stacking plies adhesive of fiber interface behavior.

IOP Conf. Series: Materials Science and Engineering **638** (2019) 012012 doi:10.1088/1757-899X/638/1/012012

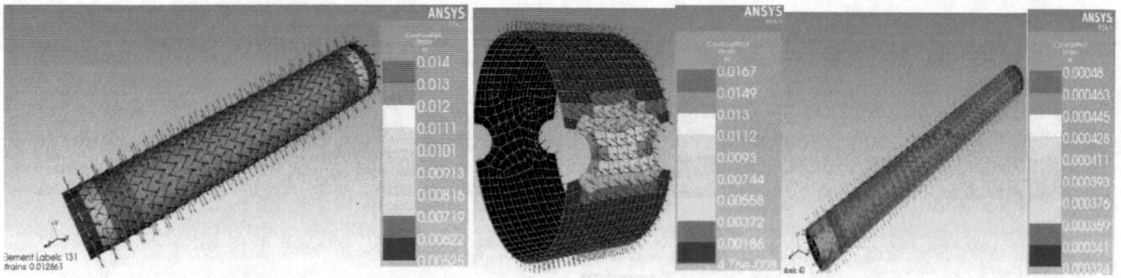

(a) (b) (c)

Figure 3. The contours of strain result for: (a) The result of hoop strain for ±55° basalt/Eglass/basalt type of composite hybrid pipe under 18 MPa biaxial internally pressure. (b) The Hoop strain of ±55° basalt/Eglass/basalt type of composite hybrid pipe under hoop tensile testing (c) Axial strain of ±55° basalt/Eglass/basalt type of composite hybrid pipe under longitudinal tensile testing

Table 3. Axial and hoop strain and stress for diverse pipe materials: (a) under 18 MPa biaxial internal pressure, (b) under 10918 N hoop tensile test and (c) under 3258 N Longitudinal tensile test

(a)

Numerical results of different materials under const. internal pressure 18 MPa

material type	Pr MPa	ε hoop	ε Long.	σ hoop MPa	σ Long MPa
3L Eglass 55	18	0.0134	-0.0082	178.03	89.8
3L Basalt 55	18	0.0132	-0.0089	178.3	89
3L BGB 55	18	0.012861	-0.0036	182.73	83
3L GBG 55	18	0.01297	-0.0038	193.8	85
3L HYBRID55	18	0.01317	-0.0021	194.23	87

(b)

Numerical Hoop tensile test results, different Materials at selected const. load

material type	Ɛh	σh	Gauge.Length mm	Max.load
3L Eglass 55	0.0181	169	9.4	10918
3L Basalt 55	0.0177	178.46	9.4	10918
3L BGB 55	0.0167	184.427	9.4	10918
3L GBG 55	0.01897	173.448	9.4	10918
3L HYBRID 55	0.0204	177.21	9.4	10918

(c)

Numerical longitudinal tens test results of different. materials at const. load

material type	ƐL	σL	A	Max.load
3L Eglass 55	0.000486	9	480	3258
3L Basalt 55	0.000467	8.9	480	3258
3L BGB 55	0.000458	9.46	480	3258
3L GBG 55	0.000463	9.42	480	3258
3L HYBRID 55	0.000466	9.07	480	3258

Form the results, it can be detected that when the stacking plies is diverse within the subjection for the same level of internal pressure value, the value of hoop strain 0.012861 mm/mm occurred for ±55° with 3 layers basalt/Eglass/basalt hybrid pipe type is lower than the rest. The hoop tensile test results indicated that the hoop strain and for ±55° with 3 layers basalt/Eglass/basalt hybrid pipe type equal to 0.0167 mm/mm under the same hoop tensile load is lower than the rest. While for longitudinal tensile test the longitudinal strain for ±55° with 3 layers basalt/Eglass/basalt hybrid pipe type equal to 0.000458 mm/mm under the same longitudinal tensile load is lower than the rest. The results in the lower strain values for the pure of basalt and Eglass with the three hybrid types within the same load condition for the three different test modes subjection indicate that the hybrid shows optimal performances than pure of basat and Eglass, also among the three hybrid pipe type, the results indicated that stacking of basalt/Eglass/basalt hybrid type shows better performance more than the rest.

And the basalt/Eglass/basalt fiber reinforced epoxy hybrid pipe can carry more internal, hoop and longitudinal tensile load earlier the tube fails. For the basalt/Eglass/basalt tube, a lower strain has been detected due to the strength of basalt fiber that has a greater elastic modulus more than Eglass. The property of an adhesive for Eglass fiber enhances the stacking for the plies interface when it occupies the middle layer among the basalt plies which will be result in better mechanical behavior. Figure.4 shows the effect of different stacking plies fiber material on the hoop strain (a) under a biaxial of an internal pressure value equal to 18.0 MPa, (b) under the hoop tensile load of 10918.0 N and (c) under a axial tensile load of 3258.0 N. It can be detected that the lower hoop strains recorded value with basalt/Eglass/basalt hybrid type. A lesser axial strains values were noted for the basalt/Eglass/basalt hybrid type also.

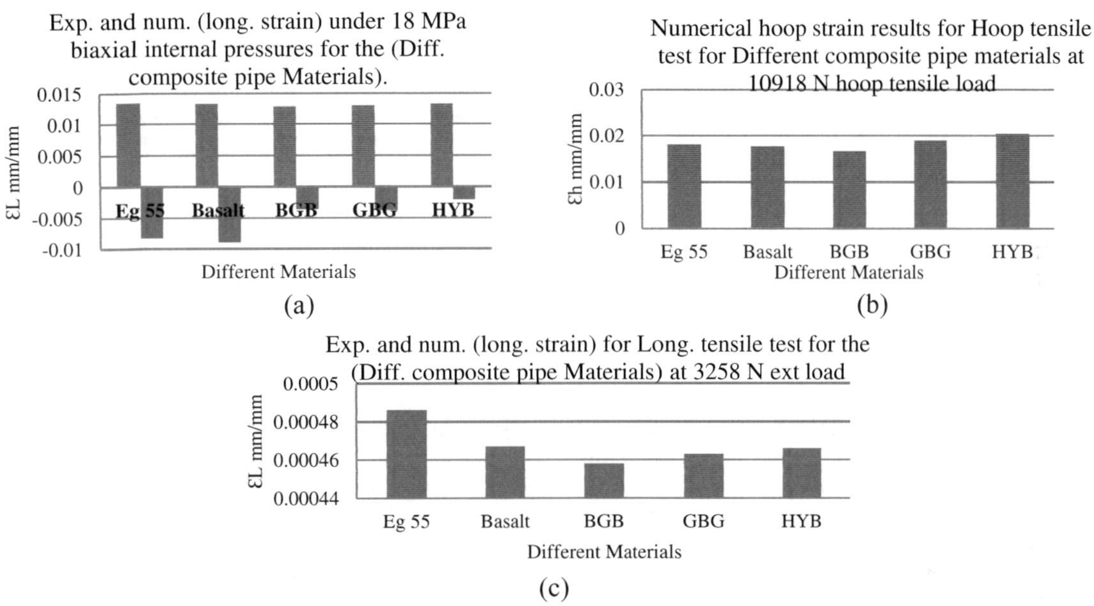

Figure 4. Effect different stacking plies fiber material on hoop and longitudinal strains (a) under 18 MPa biaxial internal pressures, (b) under 10918 N hoop tensile test and (c) under 3258 N axial tensile test

5. Conclusion

In this numerically elastic and linear stress and strain investigates of composite pipes of hybrid of Eglass and basalt fiber reinforced epoxy with different stacking sequence as well as the pure of Eglass and the pure of basalt ±55 filament wound hollow cylinder were carried out under the subjection of internal pressures as well as longitudinal and hoop tensile test. A designated model of finite element was recognized to evaluate the performance for a new structure of hybrid of basalt and Eglass reinforced epoxy over the pure of Eglass/epoxy and the pure of basalt/epoxy composite tube. Also to show the optimal stacking plies among all of hybrid different plies stacking composite pipe that has been utilized within different application. From the analysis results it concluded that:

- The basalt/epoxy showed better mechanical behavior than Eglass/epoxy composite pipe.
- Basalt/epoxy filament wound composite tube can be replaced by Eglass/epoxy tube within many healthy pressurized pipelines applications successfully as a better replacement with good mechanical performances than Eglass/epoxy.
- The hybrid basalt fiber and Eglass fiber reinforced composite pipe structure showed enhanced mechanical performances than the pure of each one.
- Basalt/Eglass/basalt fiber reinforced epoxy hybrid composite pipe has been seen as an optimal stacking plies among the three different stacking plies of hybrid composite pipe.

Based on this study, hybridization of basalt with Eglass fiber augmented the mechanical behavior, increased the mechanical strength, modulus and fatigue behavior, made costs less, excellent and has desired mechanical properties.

Acknowledgment
A gratefully acknowledged has been appreciated for the supporting of Universiti Teknologi Malaysia through the research grant No. (Q.J130000.2524.07H19).

References

[1] Fiore V, Bella GD, Valenza A. Glass-basalt/epoxy hybrid composites for marine applications. Mater Des 2011;32(4):2091–9.
[2] Dehkordi MT, Nosraty H, Shokrieh MM, Minak G, Ghelli D. Low velocity impact properties of intra-ply hybrid composites based on basalt and nylon woven fabrics. Mater Des 2010;31(8):3835–44.
[3] Sarasini F, Tirillo J, Valente M, Ferrante L, Cioffi S, Iannace S, et al. Hybrid composites based on aramid and basalt woven fabrics: impact damage modes and residual properties. Mater Des 2013;49:290–302.
[4] Saravanan D. Spinning the rocks – basalt fibres. J Inst Eng India – Part TX 2006;86:39–45.
[5] Ary Subagia IDG, Kim Y, Tijing LD, Kim CS, Shon HK. Effect of stacking sequence on the flexural properties of hybrid composites reinforced with carbon and basalt fibers. Compos Part B – Eng 2014;58:251–8.
[6] Dehkordi MT, Nosraty H, Shokrieh MM, Minak G, Ghelli D. Low velocity impact properties of intra-ply hybrid composites based on basalt and nylon woven fabrics. Mater Des 2010;31(8):3835–44.
[7] Han SH, Oh HJ, Lee HC, Su SS. The effect of post-processing of carbon fibers on the mechanical properties of epoxy-based composites. Compos Part B – Eng 2012;45(1):172–7.
[8] Park R, Jang J. Performance improvement of carbon fiber polyethylene fiber hybrid composites. J Mater Sci 1999;34(12):2903–10.
[9] Banibayat P, Patnaik A. Variability of mechanical properties of basalt fiber reinforced polymer bars manufactured by wet-layup method. Mater Des 2014;56:898–906.
[10] Carmisciano S, Rosa IMD, Sarasini F, Tamburrano A, Valente M. Basalt woven fiber reinforced vinylester composites: flexural and electrical properties. Mater Des 2011;32:337–42.
[11] Liu Q, Shaw MT, Parnas RS, McDonnell AM. Investigation of basalt fiber composite mechanical properties for applications in transportation. Polym Compos 2006;27(1):41–8.
[12] Szabo JS, Czigany T. Static fracture and failure behavior of aligned discontinuous mineral fiber reinforced polypropylene composites. Polym Test 2003;22(6):711–9.
[13] Mohanty AK., Misra M., Hinrichsen G., Biofibres biodegradable polymers and biocomposites: An overview, Macromol Mater Eng. 276/277 (2000) 1–24.
[14] Mohanty AK., Drzal LT., Misra M., Novel hybrid coupling agent as an adhesion promoter in natural fiber reinforced powder polypropylene composites, J Mater Sci Lett. 21 (2002)1885–1888.
[15] Misri S., Leman Z., Sapuan S., and Ishak M., Mechanical properties and fabrication of small boat using woven glass/sugar palm fibres reinforced unsaturated polyester hybrid composite, IOP Conference Series: Materials Science and Engineering. 11 (2010) 12-15.
[16] Das S. The cost of automotive polymer composites: a review and assessment of DOE's lightweight materials composites research. Oak Ridge National Laboratory; 2001. p. 1–49.
[17] Lopresto V, Leone C, Iorio ID. Mechanical characterisation of basalt fibre reinforced plastic. Compos Part B – Eng 2011;42(4):717–23.
[18] Manikandan V, Jappes JTW, Kumar SMS, Amuthakkannan P. Investigation of the effect of surface modifications on the mechanical properties of basalt fibre reinforced polymer composites. Compos Part B – Eng 2012;43(2):812–8.

[19] Artemenko SE, Kadykova YA. Polymer composite materials based on carbon, basalt and glass fibres. Fibre Chem 2008;40(1):37–9.
[20] Larrinaga P, Chastre C, Biscaia HC, San-Jose JT. Experimental and numerical modeling of basalt textile reinforced mortar behavior under uniaxial tensile stress. Mater Des 2014;55:66–74.
[21] Sim J, Park C, Moon DY. Characteristics of basalt fiber as a strengthening material for concrete structures. Compos Part B – Eng 2005;36(6–7):504–12.
[22] Ary Subagia IDG, Tijing LD, Kim Y, Kim CS, Vista FG-IV, Shon HK. Mechanical performance of multiscale basalt fiber–epoxy laminates containing tourmaline micro/nano particles. Compos Part B – Eng 2014;58:611–7.
[23] Foulk JD., Akin DE., Dodd RB., New low cost flax fibers for composites. SAE 2000 World Congress. (2000) 6–9.
[24] Xia M., Takayanagi H., and Kemmochi K., Analysis of multi-layered filament-wound composite pipes under internal pressure, Composite Structures. 53 (2001) 483-491.
[25] Xia M., Kemmochi K., and Takayanagi H., Analysis of filament-wound fiber-reinforced sandwich pipe under combined internal pressure and thermomechanical loading, Composite structures. 51(2001) 273-283.
[26] Rosenow M. W. K., Wind angle effects in glass fibre-reinforced polyester filament wound pipes, Composites. 15(1984), 144-152.
[27] Shultz, M., and Smith, L. V., Optimal fiber orientation for fiber reinforced pressure vessels, Applied Mechanics. (2004) 7-10.
[28] Highton,J.,Adeoye, A. B.,and Soden, P. D.,Fracture stresses for±75 degree filament wound grp tubes under biaxial loads,The Journal of Strain Analysis for Engineering Design. 20(1985)139-150.
[29] Eckold, G. C., D. Leadbetter, Lamination theory in the prediction of failure envelopes for filament wound materials subjected to biaxial loading, Composites. 9 (1978) 243-246.
[30] Carswell, W. S., An atlas of polymer damage: Lothar Engel, Herman Klingele, Gottfried W. Ehrenstein and Helmut Schaper (Eds.), Wolfe Science Books, International Journal of Fatigue 3 (1981) 218.
[31] Spencer, B. and D. Hull, Effect of winding angle on the failure of filament wound pipe. Composites, 9 (1978) 263-271.
[32] Bakaiyan H., Hosseini H., and Ameri E., Analysis of multi-layered filament-wound composite pipes under combined internal pressure and thermomechanical loading with thermal variations, Composite Structures. 88 (2009) 532–541.
[33] Wang X, Wu Z, Wu G, Zhu H, Zen F. Enhancement of basalt FRP by hybridization for long-span cable-stayed bridge. Compos Part B – Eng 2013;44(1):184–92.
[34] Zhang Y, Yu C, Chu PK, Lv F, Zhang C, Ji J, et al. Mechanical and thermal properties of basalt fiber reinforced poly(butylene succinate) composites. Mater Chem Phys 2012;133:845–9.
[35] Kunel S., A short review on Basalt fiber, International Journal of Textile Science. 1 (2012) 19-28.
[36] ASTM D2290-00, Standard Test Method for Apparent Hoop Tensile Strength of Plastic or Reinforced Plastic Pipe by Split Disk Method, ASTM International, West Conshohocken, PA, (2000), www.astm.org.
[37] ASTM D2105-01, Standard Test Method for Longitudinal Tensile Properties of "Fiberglass" (Glass-Fiber-Reinforced Thermosetting-Resin) Pipe and Tube, ASTM International, West Conshohocken, PA, (2014), www.astm.org.
[38] ASTM D1599-14, Standard Test Method for Resistance to Short-Time Hydraulic Pressure of Plastic Pipe, Tubing, and Fittings, ASTM International, West Conshohocken, PA, (2014), www.astm.org.
[39] ASTM D2584-11, Standard Test Method for Ignition Loss of Cured Reinforced Resins, ASTM International, West Conshohocken, PA, (2011), www.astm.org.
[40] Jones, M.L.C. and D. Hull, Microscopy of failure mechanisms in filament-wound pipe, Journal of Materials Science. 14 (1979) 165-174.
[41] Almula, T. A. D. M. S., M. Y. Yahya, et al., Mechanical Behavior of Functionally Graded Sandwich Hollow Cylinders under Internal Pressure, Advanced Materials Research 845 (2014) 360-366

IOP Conf. Series: Materials Science and Engineering **638** (2019) 012013 doi:10.1088/1757-899X/638/1/012013

Evaluation of the Slow Release Oxygen Properties and pH Control Ability of Oxygen Slow- releasing Materials

Zehao Chen[1], Pengfei Guo[2], Hailong Tian[1], Ying Wang[1]* and Chenghong Feng[1]

[1] The Key Laboratory of Water and Sediment Sciences, Ministry of Education, School of Environment, Beijing Normal University, Beijing, 100875, China
[2]Colleges of Chemical Engineering, Beijing University of Chemical Technology, Beijing, 100875, China

*Email: yingwang@bnu.edu.cn

Abstract. In this study, an oxygen slow-releasing materials (OSRM) was developed by simple one-step synthesis. The composition of OSRM included calcium peroxide, bentonite or carclazyte, cement, potassium dihydrogen phosphate (KH_2PO_4) and ammonium sulphate($(NH_4)_2SO_4$). The OSRM was natural and non-toxic. The slow release oxygen properties and pH control ability of OSRM were evaluated. Results showed that the oxygen concentration increased and kept stable in a week after the addition of OSRM. The OSRM added with KH_2PO_4 and $(NH_4)_2SO_4$ was better than The OSRM doped with carclazyte in controling the pH approaching to neutralization. Besides, The OSRM doped with carclazyte instead of bentonite led to the oxygen release slowly. In stimulated sediment environment, the pH value in water was stable at about 7.6 after 4 g/L of OSRM added. This phenomenon may be due to functional groups with buffer capacity occurred in sediment.

1. Introduction

The content of dissolved oxygen in water is an important parameter to measure the quality of water environment, which can also indicate the impact of human activities on sediments and water [1].For example, the eutrophication of lakes and rivers and the ability of microorganisms to degrade pollutants are often related to the level of dissolved oxygen [2].

Nowadays the aeration technology mainly includes mechanical aeration technology and chemical aeration technology. Compared with chemical oxygenation technology, mechanical aeration technology such as artificial aeration [3] is rarely used in some water bodies due to high energy consumption and low oxygenation effect. In addition, aeration is only suitable for wastewater treatment and small river channel restoration at present, and it is not convenient for large-scale aeration for lake restoration. Compared with mechanical aeration, chemical oxygenation technology mainly supply DO in water body in a simple but efficient way such as adding oxygen release materials, which could not only release O_2 to stimulate aerobic respiration of indigenous microorganisms, but also reduce the generation of anaerobic metabolites. Besides, after the addition of oxygen slow-releasing materials (OSRM) into the water body, a layer of aerobic film is formed at the sediment-water interface, which can effectively prevent the release of pollutants from the bottom mud to the overlying water body, especially the release of phosphorus in the bottom mud [4]. Therefore, OSRM is now used in different kinds of water environment restoration project such as lake restoration [5] ~ [6] and groundwater remediation [7] ~ [9].

The performance of OSRM on water restoration is affected by numerous factors, such as the compatibility of the OSRM with the environment [10] ~ [11], the internality of the OSRM [12], and its effective oxygen-releasing characteristics [13]~[14]. Traditionally, single solid peroxides such as CaO_2 are generally mixed with cement as OSRM to release oxygen upon reaction with water as in equation (1).

$$CaO_2+H_2O \rightarrow 0.5O_2+Ca(OH)_2 \quad (1)$$

Although OSRM can effectively increase DO concentration, a large proportion of oxygen escapes inevitably before entering the sediment environment, resulting in low oxygen utilization efficiency. Besides, Calcium hydroxide produced from CaO_2 released hydroxyl ion, causing high pH problem in water.

In order to minimize the unnecessary OSRM consumption and overcome the high pH problem, different oxygen slow-releasing materials(OSRM) have been developed to slow down the oxygen release properties [15]. Zhang et al. synthesized an oxygen releasing material from CaO_2, bentonite and two kinds of butters [16]. The results of evaluating performance properties of OSRM revealed that oxygen was released slowly by oxygen releasing material and the DO concentration was maintained at over 5 mg/L for a long period. Li et al. also synthesized an oxygen releasing material doped with humic acid sodium and zeolite in water restoration and found that OSRM could slow down the oxygen properties but the DO concentration was too low at 1.32 mg/L [17]. Besides, an oxygen slow-releasing material (OSRM) consisted of calcium peroxide (CaO2), stearic acid (SA) and quartz sand exhibited higher oxygen-releasing capacity, and more effective pH control due to the change SA content [18]. Otherwise, oxygen releasing alginate beads (ORABs) was synthesized by using calcium peroxide (CaO2) embedded sodium alginate is proposed to overcome the limitations regarding fast oxygen release rate and high pH induced by OSRM application [19]. However, in all of the researches above, the high pH is still big problem.

In this study, an oxygen slow-releasing material (OSRM) was synthesized. This OSRM was a mixed materials composed of calcium peroxide as oxygen applying agent, bentonite or carclazyte as slow-release-formutation, cement as binder, potassium dihydrogen phosphate (KH_2PO_4) and ammonium sulphate($(NH_4)_2SO_4$) as butters. Besides, OSRM was evaluated on the release ability of oxygen and regulation of pH through the comparison and a stimulated sediment environment was set to research the influence of bottom mud on overcoming the high pH problem caused by OSRM.

2. Materials and method

2.1. Preparation of OSRM

An oxygen slow-releasing material (OSRM) was prepared as follows. A certain percentage of calcium peroxide (CaO_2), cement, bentonite or carclazyte, potassium dihydrogen phosphate (KH_2PO_4) and ammonium sulfateere ($(NH_4)_2SO_4$) were put in a 100 mL beaker and stirred for 5-10 min. After that, a certain amount of water was added into mixture and stirred for 5-10 min. Finally, a number of cubes with a diameter of 2cm were produced by forming mixture into a mould and was placed in a ventilated and air-dried place and for 24h. The composition of different kinds of OSRCs was shown in the Table 1.

2.2. Evaluation of the oxygen-releasing properties of OSRM

An oxygen release experiment was performed in a sealed conical flask (total volume: 500 mL). Another sealed conical flask with 1cm thick layer of mud was used to test the influence of sediment.

To simulate the low level of dissolved oxygen (DO) in common water, 0.3 g /L of Na_2SO_3 was added to both sampling to maintain the DO concentration below 0.1 mg/L. the pH of water was adjusted to neutral. The oxygen concentration was measured using a DO meter (JPBJ-608, China). The pH was measured using a pH meter (PHSJ4F, China).The testing time is not more than one week. The dosage of OSRM was about 4g/L for ORSM -A, ORSM-B, ORSM-B, ORSM-C and ORSM-D. The dosage of OSRM was about 2g/L for ORSM -E, ORSM-F and ORSM-G.

Table 1. The composition of different kinds of OSRM

Content (g/4gORSM)	A	B	C	D	E	F	G
CaO_2	1	1	1	1	2	2	2
Cement	0.5	0.5	0.5	0.5	1	1	0.5
bentonite	0.5	0.5	0.5	0.5	1	0	0
carclazyte	0	0	0	0	0	1	1.5
KH_2PO_4	2	0.5	0.25	0.125	0	0	0
$(NH_4)_2SO_4$	0	1.5	1.75	1.875	0	0	0

3. Result and discussion

3.1. The Evaluation of oxygen-releasing properties and pH control ability for OSRM

The dissolved oxygen concentrations and pH changes of the OSRM with different content buffer chemicals were shown in Fig.1 (a). In the first day, oxygen concentrations increased obviously and then become stable between 5.5mg/L and 7.2 mg/L. Wherein, the dissolved oxygen concentrations released by OSRM-B, OSRM-C and OSRM-D are similar and more stable.

The pH changes were shown in Fig.1 (b). The highest pH value was observed for OSRM-A. The addition of ammonium sulfate was benefit for the pH control and also helped the concentration of O_2 maintain at about 6.5 mg/L. So the addition of KH_2PO4 and $(NH_4)_2SO4$ as buffer chemicals could control the pH to stable at neutralization.

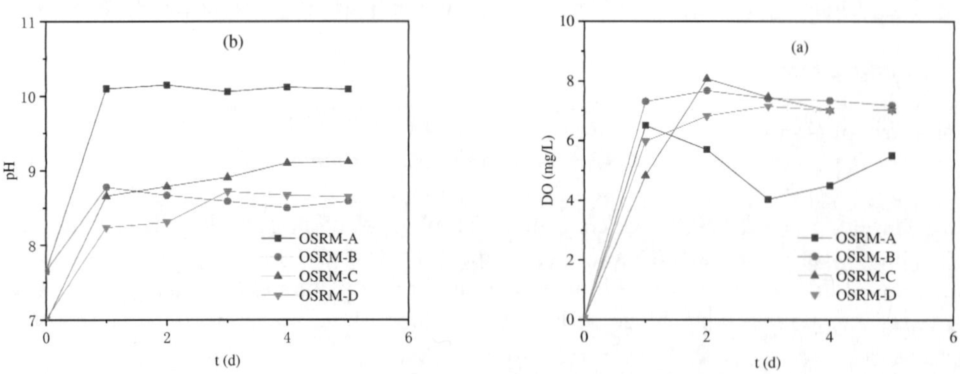

Figure 1. The effects of KH_2PO_4 and $(NH_4)_2SO_4$content of OSRM on DO concentration (a) and pH (b)

3.2. The Effects of doped carclazyte and bentonite on release of oxygen and regulation of pH

The effect of doped carclazyte and bentonite on pH in water and oxygen concentration was shown in Fig.2. When the content of carclazyte or bentonite was 1 g, oxygen concentrations increased obviously in the first day. Then oxygen concentrations were about 5.5 mg/L and 4.62 mg/L for OSRM-E and SRM-F, respectively. The doped carclazyte lead to the oxygen release slowly and lower pH value. However, when the increase of carclazyte contented from 0.5 to 1 g, the material was too loose to stability. Moreover, the pH value is the highest for OSRM-G.

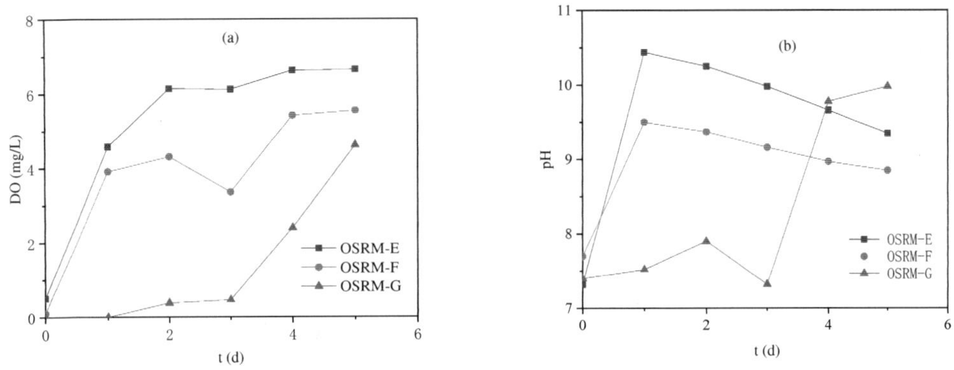

Figure 2. The effects of carclazyte and bentonite content of OSRM on DO concentration (a) and pH (b)

3.3. The Effects of simulated sediment environment on release of oxygen and pH control

In simulated sediment environment, the pH value in water (Fig. 3(a)) maintained at about 7.6 by the addition of SRM-A in the whole 6 days, which may be due to functional groups with buffer capacity occurred in sediment. Comparably, the pH value in water increased sharply in the first day without sediment, then stable at about 10.0. The concentration of O2 (Fig. 3(b)) rose sharply in the first day and then decreased slowly. The concentration of O_2 was similar, which maintained between 4.5 mg/L and 6.5 mg/L.

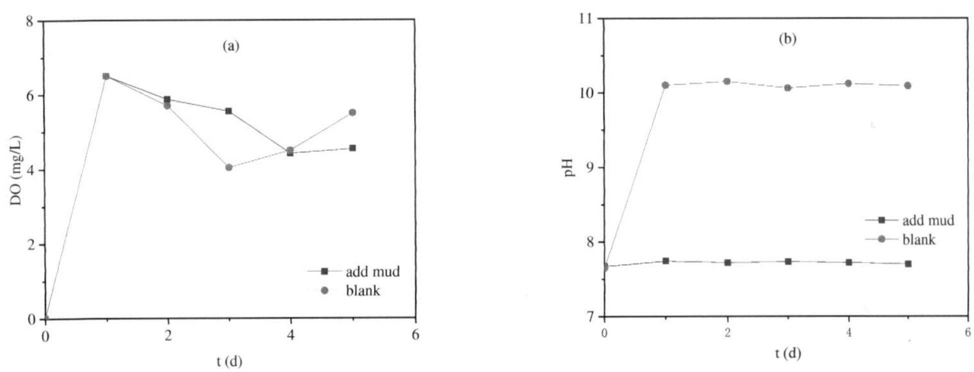

Figure 3. The Effects of simulated sediment environment on DO concentration (a) and pH(b)

4. Conclusions

In this study, slow release oxygen materials were developed. The slow release oxygen properties and pH control ability of oxygen slow-releasing material were evaluated. Oxygen concentration was increased and kept stable in a week. The addition of KH_2PO_4 and $(NH_4)_2SO_4$ as buffer chemicals may control the pH to become stable at neutralization. The doped carclazyte instead of bentonite led to the oxygen release slowly and lower pH value. In simulated sediment environment, the pH value in water stable at about 7.6, which may be due to functional groups with buffer capacity occurred in sediment.

In the future, the effect of oxygen release materials on pollutants such as nitrogen and phosphorus release from sediment should be further studied.

Acknowledgments
We greatly acknowledged the Major Science and Technology Program for Water Pollution Control and Treatment (2018ZX07110004).

Reference

[1] Wang J J, ShenJ, Zhang L, et al. Determination of oxygen exchange rate at lake sediment-water interface and its influencing factors [J]. Lake science, 2009. 21(4): 474-482.
[2] Li H, Song C L, Cao, X Y., Zhou, Y Y. The phosphorus release pathways and their mechanisms driven by organic carbon andnitrogen in sediments of eutrophic shallow lakes. Sci. Total Environ, 2016.572, 280–288.
[3] Freeman A I, Surridge B W J , Matthews M , Stewart M , Haygarth P M. New approaches to enhance pollutant removal in artificially aerated wastewater treatment systems. Sci. Total Environ. 2018. 627.
[4] Zhou J, Li D, Chen S, et al. Sedimentary phosphorus immobilization with the addition of amended calcium peroxide material[J]. Chemical Engineering Journal, 2019.357:288-297..
[5] Thiruvenkatachari R, Vigneswaran S, Naidu R. Permeable reactive barrier for groundwater remediation [J]. Ind. Eng. Chem. 2008,14,145-156
[6] Wang Y, Wang W H, Yan F L, et al. Effects and mechanisms of calcium peroxide on purification of severely eutrophic water[J]. Sci. Total Environ., 2019, 650:2796-2806.
[7] Cao X, Harris W. Properties of dairy-manure-derived biochar pertinent to its potential use in remediation. Bioresour. Technol. 2010,101, 5222-5228.
[8] Mosmeri H, Tasharrofi S, Alaie E, et al. Controlled-release oxygen nanocomposite for bioremediation of benzene contaminated groundwater[J]. New Polymer Nanocomposites for Environmental Remediation, 2018:601-622.
[9] Lu, S., Zhang, X., Xue, Y., 2017. Application of calcium peroxide in water and soil treatment: a review. J. Hazard. Mater. 337, 163.
[10] Severino P, Chaud M V, Shimojo A, Antonini D, Lancelloti, M, Santana M H A, Souto E B.. Sodium alginate-cross-linked polymyxin B sulphate-loaded solid lipid nanoparticles: antibiotic resistance tests and HaCat and NIH/3T3 cell viability studies. Colloids Surf. B. 2015,129, 191-197.
[11] Meng, F., Zhong, Z., Feijen, J.Stimuli-responsive polymersomes for programmed drug delivery. Biomacromolecules 2009,10, 197-209.
[12] Nykanen A, Kontio H, Klutas O, Penttinen O P, Kostia S, Mikola J, Romantschuk M. Increasing lake water and sediment oxygen levels using slow release peroxide. Sci. Total Environ. 2012.429, 17-24.
[13] Qian Y, Zhou, X, Zhang Y, Zhang W, Chen J. Performance and properties of nanoscale calcium peroxide for toluene removal. Chemosphere.2013, 91, 717-723.
[14] Ahmad F, Schnitker S P, Newell, C J. Remediation of RDX and HMX contaminated groundwater using organic mulch permeable reactive barriers. J. Contam. Hydrol. 2007, 90, 1-20.
[15] Daghighi S, Sjollema J, van der Mei, H C, Busscher, H J, Rochford, E T J. Infection resistance of degradable versus non-degradable biomaterials: an assessment of the potential mechanisms. Biomaterials. 2013, 34, 8013-8017.
[16] Zhang YX, Liang JK, Wang R, et al. Application of improving oxygen releasing material in situ polluted groundwater PRB bioremediation. Chinese Journal of Environmental Engineering. 2012,6:3910-3914.
[17] Li H, Wu C Y et al. Preparation of composite oxygen release agent and its effect on water restoration [J]. Environmental engineering, 2017(9).
[18] Zhou Y, Fang X, Zhang Z, et al. An oxygen slow-releasing material and its application in water remediation as oxygen supplier[J]. Environmental Technology, 2017:1-7.

[19] Lee C-S, Le Thanh T, Kim E-J, et al. Fabrication of novel oxygen-releasing alginate beads as an efficient oxygen carrier for the enhancement of aerobic bioremediation of 1,4-dioxane contaminated groundwater. Bioresour. Technol. 2014, 171:59-65.

IOP Conf. Series: Materials Science and Engineering **638** (2019) 012014 doi:10.1088/1757-899X/638/1/012014

Agri-residual Filler Based Bio- Composites: *Synergistic Effect of Organo-sulfide Coupling Agent and Alkali Treatment on Mechanical Properties*

Firoz Alam Faroque[1*], Saanil Kakati[1], Sanchita Bandhopadhyay-Ghosh[2] and Subrata Bandhu Ghosh[2]

1 Department of Civil Engineering, Manipal University Jaipur, Jaipur, India-303007
2 Department of Mechanical Engineering, Manipal University Jaipur, Jaipur, India-303007

*E-mail: firozalam.faroque@jaipur.manipal.edu

Abstract. With growing consumer awareness, carbon footprint and waste disposal challenges, there is an increasing demand for the use of agricultural wastes in developing value-added structural components. In this current study, commercially available epoxy resin has been reinforced with chemically modified wheat straw filler. Particulate wheat straw fillers were undergone alkaline treatment to remove selective and non-essential components. Alkaline modified samples were then treated with organo-sulfide based coupling agent to enhance filler/matrix interfacial compatibility. The bio-composite samples were developed using casting technique. The cast samples were tested for their mechanical strengths using compression and flexural tests. The results indicated that the use of coupling agent and alkaline treatment led to improvement of mechanical strength in a synergistic manner. Optical micrographs of fractured surfaces also revealed morphological features which could be correlated to the mechanical performance of the bio-composites.

1.Introduction

With the growing population, the demand for food crops is ever growing, thus leading higher production agricultural crops. Rice, wheat, corn are some the major agricultural crops worldwide. Looking into Indian scenario, a big part of Indian economy depends on agricultural production. Annually around 500-550 million tonnes (Mt) of crop residue is generated in India of which majority is wheat and rice. Even after cattle feed, animal bedding, cooking fuel, organic manure, more than 30 percent of the total residue remain unused [1]. Waste management of such a large quantity of unused wheat straw is a very big challenge. It is often found that farmers have no option but to burn it which contributes to air pollution. Thus, for sustainability reasons this huge amount of agricultural waste are being used in manufacturing greener products for different applications. Agricultural waste is available abundantly and it is renewable resource. As the oil resources are depleting, researchers are shifting towards green/natural resources because of social and environmental concern [2]. Use of natural fibre in manufacturing of composite in various industries has been there for some time as agricultural waste has shown potential as filler in the polymer matrix be it thermoset or thermoplastic polymer [3-4]. Natural fibres are slowly becoming an alternative to glass fibre and carbon fibre which are very popular for their strength and other advantageous properties. A varied applications of bio-composites can be found in construction, automobile, packaging and medical industry[5].

Thus, the use of agricultural waste as a filler in the polymer matrix to obtain composite material of desirable properties are becoming more and more popular among researcher. However, working with raw natural fibre is challenging because of their hydrophilicity and in turn the incompatibility with traditional polymer matrices which are hydrophobic in nature. Because of this nature, the bonding between fibre and matrix becomes weak and affects the resultant mechanical properties of the composites. Raw fibres must undergo some kind of physical or chemical modification before being used in polymer matrix in order to get better mechanical properties [6], [7]. Because of abundance and renewability, working with wheat straw fibres as filler in polymer matrix is popular option among researchers and scientist. It is found in literature that researchers have used wheat straw and wheat straw blended with other natural fibres as a filler material to reinforce polymer matrix and have reported that wheat straw fibres are a good alternative to manufacture green composite material and its application in various domains [8], [9].

In this paper, therefore we have investigated the combined effect of organo-sulfide Si 69 coupling agent and alkali treatment on the mechanical performance of wheat straw particulate filer reinforced bio-composites.

2.Materials

Epoxy resin (SMR 2085) of epoxide number 4.93 and viscosity of 3200 cPs at 25°C has been is catalysed by triethylenetetramine ($C_6H_{18}N_4$) hardener of density 0.980 g/cc at 20°C and refractive index of 1.499 hardener has been used. An organo-sulfide based coupling agent Si 69 has been used in the study. Wheat straw was obtained from local agricultural field. All chemical reagents were of laboratory grade.

3. Methods

3.1 Pre-treatment of wheat straw

Wheat straw was collected from the local agricultural field in Jaipur for this research. The wheat straw fibres were converted into powder form using a standard mixer. This sample was then passed through the sieve 450-micron sieve and the portion that passed through the sieve was used as filler in the epoxy matrix. This powdered wheat straw fibres were then kept in 5% NaOH solution for 24 hours. It was then washed properly using distilled water. The washed wheat straw filler was then oven dried for 24 hours at 60°C temparature. The alkali treated wheat straw fibres were then modified with organo-sulfide coupling agent (Si 69). These modified fibres were washed again with distilled water and then oven dried for 24 hours at 60° C for 24 hours before using it in the epoxy matrix.

3.2 Fabrication of bio-composites

Milled wheat straw particulates were then used in proportions of 10% and 15% by weight of epoxy as a filler material in the epoxy matrix. The ingredients were then mixed using a mechanical stirrer (Remi motors, RQ-124A) at 500 rpm to form a uniform mixture. For further uniform distribution of the filler materials the mixture was then put in ultrasonicator for 30 minutes at room temperature. This composition was then poured into a silicon mould and was left to harden. The average hardening time was 4 hours for all samples. A, B, C,D,E,F and O are the seven different samples taken in this study. Table 1 shows the composition of this samples and the type of treatment it has been subjected to.

IOP Conf. Series: Materials Science and Engineering **638** (2019) 012014 doi:10.1088/1757-899X/638/1/012014

Table 1: Composition of Bio-composites

Sample	*Epoxy Resin*	*Wheat straw(phr*)*	*Silane treatment*	*NaOH treatment*
O	100	0	No	No
A	100	10	No	Yes
B	100	15	No	Yes
C	100	10	Yes	Yes
D	100	15	Yes	Yes
E	100	10	No	No
F	100	15	No	No

3.3 Mechanical testing

Mechanical strength plays an important parameter for any composite material be it flexural strength or compressive strength. For determination of mechanical properties, rectangular samples were cut from the prepared wheat straw epoxy composite in size of 50x10x4 for compression testing (as per ISO 604) and in size of 80x10x4 for flexural testing (according to ISO 178). At least 5 samples of each composition were tested using universal testing machine (UTM) of Dak Systems using 10 kN load cell.

3.4 Optical microscopy

Moticam 5.0 MP optical microscope has been used to observe the prepared samples. A magnification of 500x was used to see the morphology of the composite samples prepared. All the fractured samples are grinded using coarser grinder paper to finer grinder paper and then it was polished using diamond paste in the polisher to get a smooth surface to observe under the microscope.

4. Result and discussion

4.1 Mechanical properties

Figure 1 and 2 show compressive strengths and flexural strengths of the bio-composites respectively. At least five samples of each compositions were tested. It was observed that pure epoxy samples have the highest compressive and flexural strength of 45.83 MPa and 58.78 MPa respectively. This could be attributed to the fact that wheat straw fillers being hydrophilic in nature, interfacial compatibility with hydrophobic epoxy matrix would be limited. However, it was observed that effect of alkali treatment could improve both compressive and flexural strength for both fibre loading (10phr and 15 phr). The strengths were further enhanced on treatment with organo-sulfide coupling agent, indicating a synergistic effect. The much-improved dispersion of the bio-fibres and the interaction between the fibre and epoxy matrix are in good agreements with the optical micrographs as shown in figure 3. The results could be attributed to the fact that effect of alkali treatment could remove the undesirable components such as waxes and gums, thereby facilitating resin impregnation and better fibre wetting. This in turn resulted into improved mechanical properties. The effect of coupling agent could be attributed to improved interfacial compatibility between filler and epoxy matrix, thereby, resulting into enhanced mechanical strength. It was also observed from fig.1 and fig.2 that incorporation of 10 phr filler could result into optimized mechanical properties, beyond which there were reduction in mechanical strengths. This could possibly be attributed to non-uniform dispersion wheat straw fillers at higher (15 phr) loading.

IOP Conf. Series: Materials Science and Engineering **638** (2019) 012014 doi:10.1088/1757-899X/638/1/012014

4.2 Optical microscopy

Optical micrographs of the fractured surfaces of all the samples taken in this study are shown in fig.3.The figure (g) represents the fractured surface of pure epoxy sample. No voids can be observed and the surface morphology is smooth and very uniform. Micrographs (e) and (f) represents the samples prepared from raw straw and epoxy. It is clearly visible from these micrographs that there is neither proper dispersion of the bio-fibres in the polymer matrix nor proper interaction between fibre and matrix. Moreover, voids and cracks can be found in these composites. With the alkali treatment there is improvement of the interaction between fibre and matrix as observed in micrographs (a) and (b). It is because alkali treatments remove the outer layer from the wheat straw and makes the surface of the bio-fibres rough, helps in better bonding between the polymer matrix and fibre. It is also evident that the use of coupling agent resulted in enhanced fibre/matrix interfacial compatibility (micrographs c and d) which could prevent major crack formation and development of voids. Micrographs show that all the specimens failed in brittle manner. These observations are in very good agreement with the results obtained from mechanical tests.

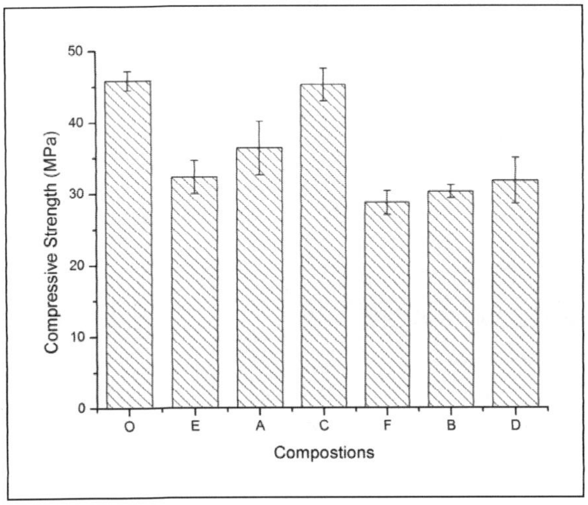

Figure 1. Compressive Strengths of the bio-composites

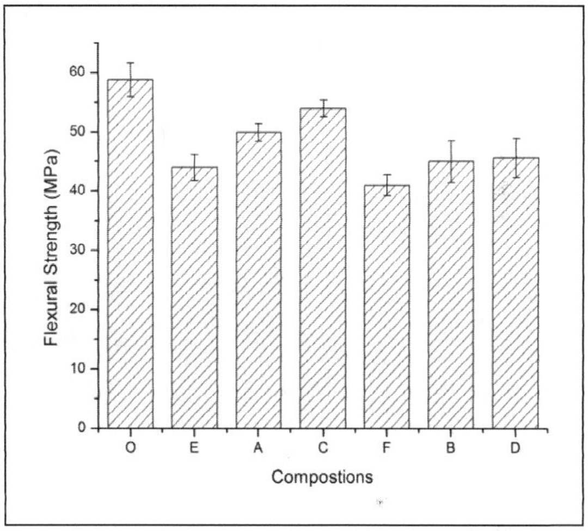

Figure 2. Flexural Strengths of the bio composites

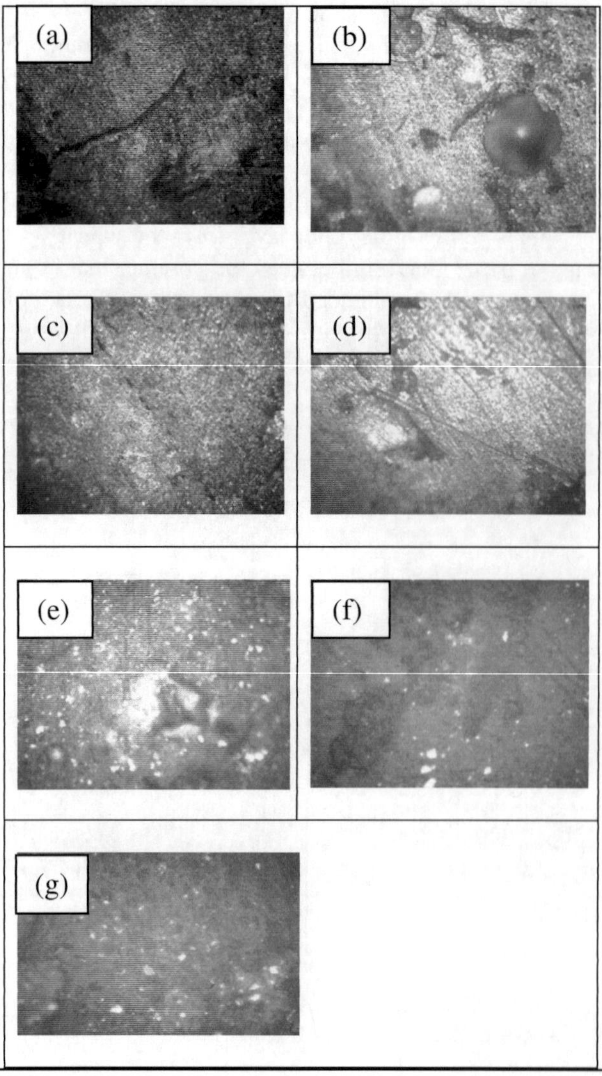

Figure 3. Micrographs of fractured samples (a) A (b) B (c) C (d) D (e) E (f) F (g) O

5. Conclusions

A range of bio-composites were developed using agri-residue based fillers (wheat straw) as reinforcements within epoxy matrix. The fillers were treated with alkali solution and then with organo-sulfide coupling agent. Although there were reductions in mechanical properties (compressive strength and flexural strength) on the incorporation of wheat straw particulate fillers compared to the pure epoxy samples but the properties were found to be enhanced on treatment with alkali solution and further enhanced by organo-sulfide coupling agent treatment. Results also indicated an optimized filler loading (10 phr) both for compressive and flexural strengths. With the increase of loading of wheat straw, there was reduction in the mechanical strengths. The fact that chemical treatments (both alkali and silane treatment) play a significant role in improving the mechanical properties because of improvement of interfacial compatibility between fibre and polymer matrix is also backed by optical micrographs showing the surface morphology of the fractured samples. Major crack formation and voids were found restricted in the coupling agent treated samples and were in good agreements with the results obtained from mechanical tests. Over all the mechanical strength of the double treated fibre specimen showed significant improvement over single treated fibres and raw fibre epoxy composite.

Funding
This work is a part of an on-going research. The authors would like to thank the management of Manipal University Jaipur, Rajasthan, India for providing the financial support for carrying out this research.

Acknowledgement
Authors would like to thank the department of mechanical and department of civil engineering, Manipal University Jaipur for providing the laboratory facility for carrying the research.

References

[1] S. Devi, C. Gupta, S. L. Jat, and M. S. Parmar, "Crop residue recycling for economic and environmental sustainability: The case of India," *Open Agric.*, vol. 2, no. 1, pp. 486–494, Sep. 2017.

[2] G. Bogoeva-Gaceva *et al.*, "Natural fiber eco-composites," *Polym. Compos.*, vol. 28, no. 1, pp. i. 98–107, Feb. 2007.

[3] E. Zini and M. Scandola, "Green composites: An overview," *Polym. Compos.*, vol. 32, no. 12, pp. 1905–1915, Dec. 2011.

[4] E. Jayamani, S. K. Heng, P. L. N. Khui, and M. K. Bin Bakri, "Comparative Study of Compressive Strength of Epoxy Based Bio-Composites," *Key Eng. Mater.*, vol. 775, pp. 68–73, 2018.

[5] K. N. Bharath and S. Basavarajappa, "Applications of biocomposite materials based on natural fibers from renewable resources: a review," *Sci. Eng. Compos. Mater.*, vol. 23, no. 2, Jan. 2016.

[6] M. J. John and R. D. Anandjiwala, "Recent developments in chemical modification and characterization of natural fiber-reinforced composites," *Polymer Composites*, vol. 29, no. 2. pp. 187–207, Feb-2008.

[7] X. Li, L. G. Tabil, and S. Panigrahi, "Chemical treatments of natural fiber for use in natural fiber-reinforced composites: A review," *Journal of Polymers and the Environment*, vol. 15, no. 1. pp. 25–33, 17-Feb-2007.

[8] V. Mittal and S. Sinha, "Study the effect of fiber loading and alkali treatment on the mechanical and water absorption properties of wheat straw fiber-reinforced epoxy composites," *Sci. Eng. Compos. Mater.*, vol. 24, no. 5, pp. 731–738, 2017.

[9] V. Mittal and S. Sinha, "Mechanical, thermal, and water absorption properties of wheat straw/bagasse-reinforced epoxy blended composites," *Adv. Polym. Technol.*, vol. 37, no. 7, pp. 2497–2503, 2018.

Optimization of Process Parameters for Closed-die Forging Based on Finite Element Simulation

Wei Guo *, Jin Zou
Institute of Applied Physics, Jiangxi Academy of Sciences, Nanchang 330096, China
*E-mail: guowei053@163.com

Abstract. In this study, finite element software was used to simulate and optimize the process parameters including billet thickness, forging speed, and friction coefficient during closed-die forging. The results show that the mean equivalent strain of the billet gradually increases as the thickness reduces. However, the corners of the billet show certain arcs and the materials are insufficiently filled when the thickness is too small. The higher the die-forging speed is, the more significantly the temperature rises and the more non-uniformly the temperature distributes in the billet. In addition, the mean equivalent strain shows little change when the friction coefficient increased from 0.1 to 0.3, but the strain distribution became more inhomogeneous.

1. Introduction

The application of severe plastic deformation (SPD) process to prepare ultrafine-grained materials has been paid more attention by researchers and witnessed rapid development in recent years[1]. There are numerous factors influencing the machining effect of SPD, such as die structure, machining path, processing temperature, processing speed, and friction coefficient between materials and die. It generally calls for huge manpower and materials to systematically study the influences of different process technologies on the materials through experiments[2-4]. With the rapid development of computer technologies and finite element software, it has provided a simple approach for analyzing the SPD process and optimizing process parameters by establishing models to carry out numerical simulation. At present, the finite element simulation of SPD methods such as equal channel angular extrusion, high pressure torsion, accumulative roll bonding and reciprocating extrusion has gained considerable research achievements[5-6]. It has been reported that cyclic closed-die forging has significant effects in refining microstructures and improving mechanical properties of materials[7-8]. However, numerical simulation is still absent for the technology. The research applied the numerical simulation software to carry out finite element analysis and optimize the parameters of closed-die forging, thus providing references for the practical application of the technology.

2. Research Procedures

A three-dimensional model was built at first. Fig.1a and Fig.1b separately illustrate the sketch map and finite element mesh generation of the die-forging device. The dimensions of the billet in Fig.1a is $100 \times 100 \times 20$ mm^3. The device consists of a punch with dimensions of $100 \times 100 \times 140$ mm^3 for external loading and a lower die with external dimensions of $180 \times 180 \times 165$ mm^3. The chamber of the lower die and the punch are of equal dimensions. The material model used for the die-forging billet was the built-in Al-1100 of database of the software and defined as a plastic body. Its material property parameters are as follows: Young's modulus is 68900 MPa, Poisson's ratio is 0.33, and thermal expansion coefficient is 2.2×10^{-5}/ K. The dies were made of H-13 steel and defined as a rigid body. The initial temperature of the dies was set as 400 °C. The mesh of the billet was divided using 9340 3-node isoparametric elements which were automatically generated by the software. Mesh generation was also conducted for the die so as to analyze its heat transfer with the billet. Nevertheless, as the plastic deformation of the die is ignored, the mesh of the die is generated with large elements. The adaptive mesh regeneration was performed as the numerical simulation of the process proceeded. A shear friction model was adopted for the friction between the billet and the die. In the post-processing module, the distributions of internal strain and temperature of the die-forged billets were obtained and displayed using nephogram and histograms.

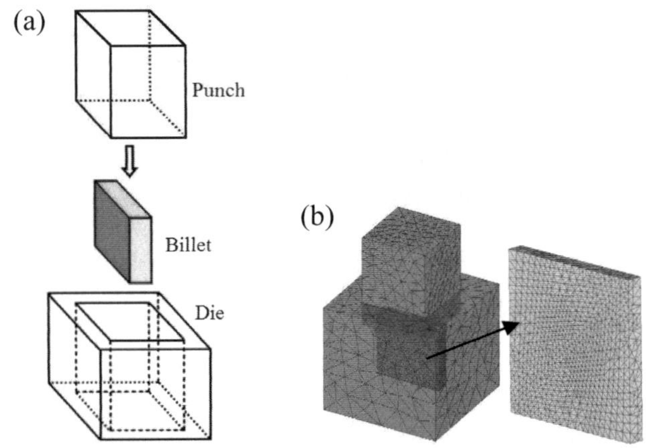

Figure 1. (a) Sketch map and (b) finite element mesh generation of closed-die forging device

3. Results and Analyses

3.1. Thickness Optimization

When the cross-sectional area of the billet is fixed, the thickness determines the reduction of die forging. The thinner the billet is, the larger the reduction. Three thicknesses, 10 mm, 20 mm, and 40 mm, were selected to carry out simulation analysis. The shapes and strain distributions after die forging under the above three thicknesses are shown in Fig.2a, Fig.2b, and Fig.2c. At the edges of the billets appear some elements with maximum value of the equivalent strain, due to the very small gap between the punch and the lower die. As illustrated in the figures, as the thickness increased from 10 mm to 20 mm and finally to 40 mm, the mean equivalent strain in the billet reduced from 2.68 to 1.99 and 1.28 eventually. The phenomenon indicates that a small thickness is able to obtain large strain and allows one to attain finer microstructure in practical machining. However, the corners of the billet showed certain arcs and the materials were insufficiently filled, and fracture would easily occur in the billet due to large strain when the thickness was 10 mm. While the mean equivalent strain was too low if the thickness of billet was 40 mm. Therefore, it is reasonable to set the thickness of the billet as 20 mm in practice.

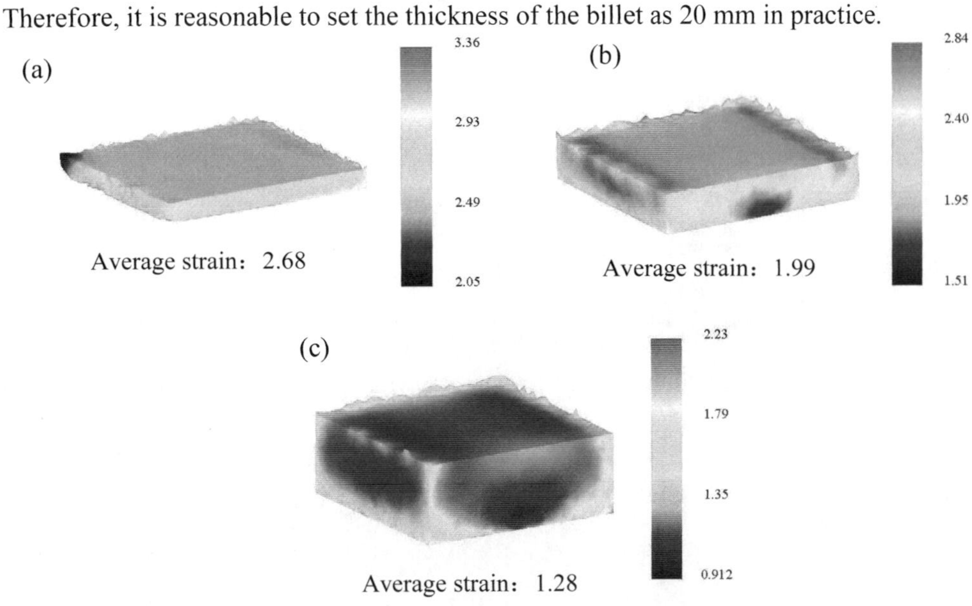

Figure 2. The shapes and strain distributions after die forging under the thicknesses (a) 10 mm, (b) 20 mm, and (c) 40 mm

IOP Conf. Series: Materials Science and Engineering **638** (2019) 012015 doi:10.1088/1757-899X/638/1/012015

3.2.Speed Optimization

The initial temperature of the billet was set as 400 °C to carry out the die-forging at the speeds of 3 mm/s, 30 mm/s, and 300 mm/s. The obtained temperature fields are illustrated in Fig.3a, Fig.3b, and Fig.3c. It can be seen from the figures that the billets presented remarkable differences in temperature rise at different die-forging speeds. The average temperatures of the billets at the die-forging speeds of 3 mm/s, 30 mm/s, and 300 mm/s were 411 °C, 424 °C, and 445 °C, with the standard deviations of 1.87 °C, 6.21 °C, and 16.4 °C, respectively. In addition, high temperature (above 480°C) was also measured at local area when the speed was 300 mm/s. The results imply that the higher the die-forging speed is, the more significantly the temperature rises and the more non-uniformly the temperature distributes. Therefore, a relatively low speed should be applied in the die forging, to avoid the overgrowth of recrystallized grains due to the excessive temperature rise of the billet.

(a)

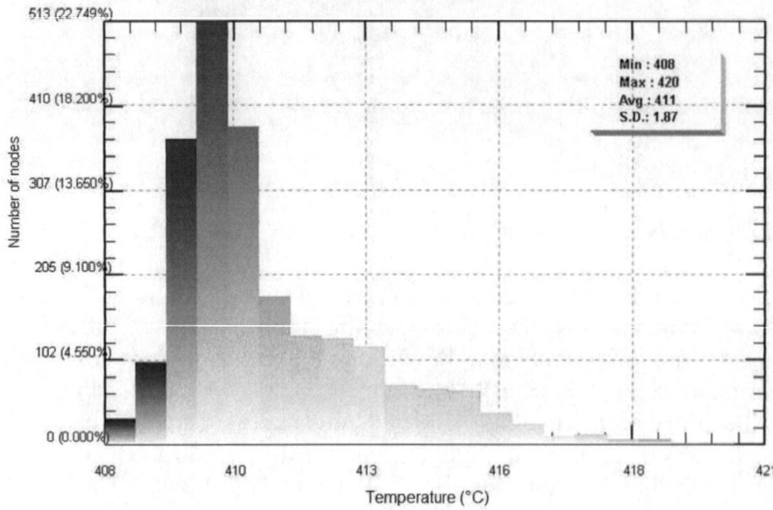

(b)

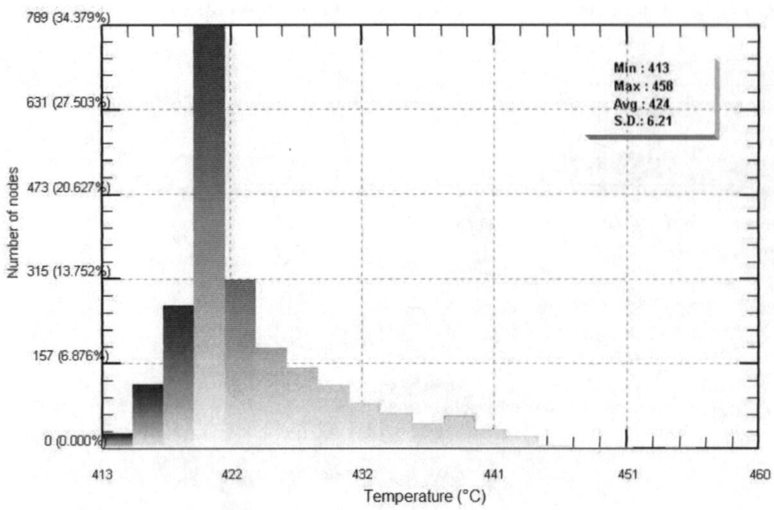

(c)

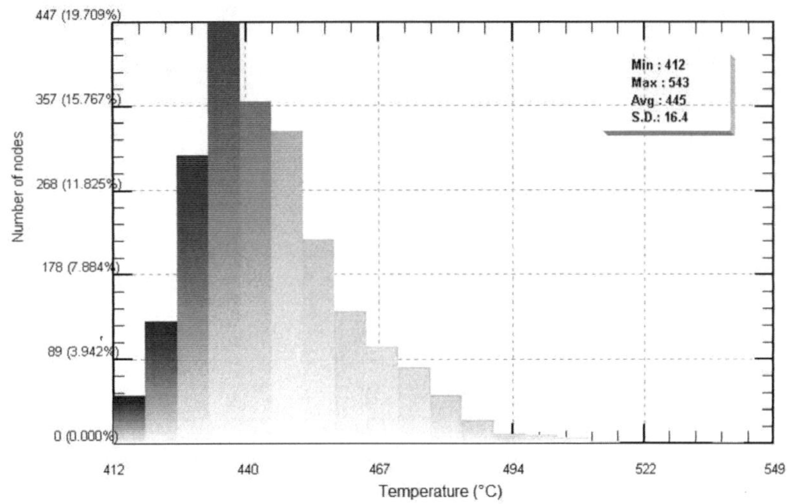

Figure 3. The temperature distributions in the billets after die forging at the speeds of (a) 3 mm/s, (b) 30 mm/s, and (c) 300 mm/s.

3.3. Optimization of Friction Coefficient

Closed-die forging was conducted with friction coefficients of 0.1 and 0.3 at the speed of 3 mm/s and temperature of 400 °C. The obtained equivalent strain distributions are demonstrated in Fig.4a and Fig.4b. The mean equivalent strain in the billet was 1.99 with a standard deviation of 0.165 when the friction coefficient was 0.1. As the friction coefficient grew to 0.3, the mean equivalent strain increased to 2.03 with a standard deviation of 0.172. The mean equivalent strain showed little change when the friction coefficient increased from 0.1 to 0.3, but the strain distribution became more inhomogeneous. Thus, lubricant should be used to reduce the friction between the billets and die in practice.

(a)

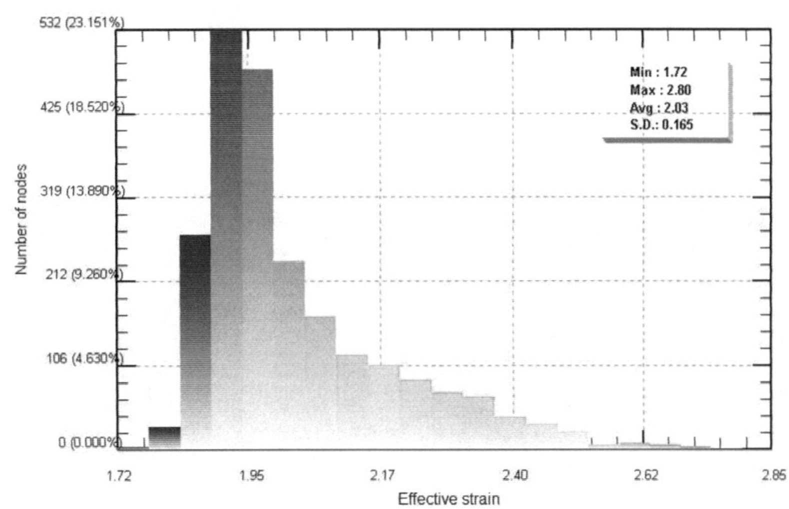

(b)

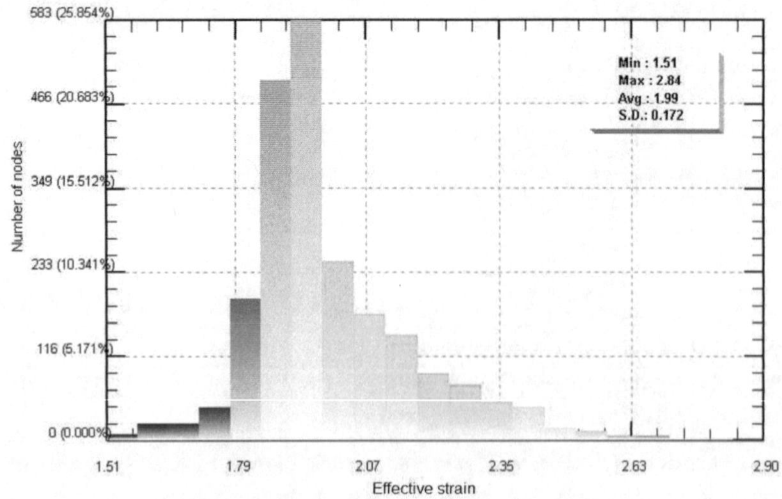

Figure 4. The strain distributions in the billets after die forging with friction coefficients of (a) 0.1 and (b) 0.3

4. Summary

In this study, finite element software was used to optimize the process parameters of billet thickness, processing speed, and friction coefficient during closed-die forging. The main conclusions are as follows:

(1) The mean equivalent strain in the billet increases from 1.28 to 2.68 as the thickness reduces from 40 mm to 10 mm. However, the corners of the billet shows certain arcs and the materials are insufficiently filled when the thickness is 10 mm. While the mean equivalent strain is too low if the thickness of billet is 40 mm. Therefore, it is reasonable to set the thickness of the billet as 20 mm in practice.

(2) As the die-forging speed increases from 3 mm/s to 300 mm/s, the temperature rise increases from 11 °C to 45 °C, and inhomogeneity of temperature distribution enhances in the billet. Therefore, a relatively low speed should be applied in the die forging.

(3) The mean equivalent strain shows little change when the friction coefficient increased from 0.1 to 0.3, but the strain distribution became more inhomogeneous. Thus, lubricant should be used to reduce the friction between the billets and die in practice.

Acknowledgments

The present research was sponsored by Key Scientific Research Project from Jiangxi Academy of Sciences under Grant no. 2017-YZD2-20, Scientific Research and Development Foundation from Jiangxi Academy of Sciences under Grant nos. 2018-YCXY-11, 2018-YYB-08, 2015–YYB–11 and 2015–XTPH1–11.

References

[1] Valiev R Z, Langdon T G, Prog. Materi. Sci. Vol. **51** (2006) pp. 881-981.
[2] Inoue T, Yanagida A, Yanagimoto J, Mater. Lett. Vol. **106** (2013) pp. 37-40.
[3] Inoue T, Tsuji N, Comp. Mater. Sci. Vol. **46** (2009) pp.261-266.
[4] Lin J B, Wang Q D, Liu M P, Chen Y J, Roven H J, Trans. Nonferrous Met. Soc. China Vol. **22** (2012) pp.1902-1906.
[5] Wei W, Zhang W, Wei K X, Zhong Y, Cheng G, Hu J, Mater. Sci. Eng. A Vol. **516** (2009) pp.111-118.
[6] Yoon S C, Horita Z, Kim H S, J. Mater. Process. Technol. Vol. **201** (2008) pp.32-36.
[7] Guo W, Wang Q D, Ye B, Zhou H, J. Alloy Compd. Vol. **558** (2013) pp.164-171.
[8] Guo W, Wang Q D, Ye B, Zhou H, J. Alloy Compd. Vol. **552** (2013) pp.409-417.

IOP Conf. Series: Materials Science and Engineering **638** (2019) 012016 doi:10.1088/1757-899X/638/1/012016

Analysis of optimal profile modification amount of spur gear under disturbance of backlash and center distance

Enen Shen, Yumei Hu, Zao He, Huan Zhang and Yanghan Min

State Key Laboratory of Mechanical Transmission, Chongqing University, Chongqing 400044, China
E-mail: enenshen@163.com

Abstract. Under high temperature conditions, backlash and center distance of gear pair deviate from the preset value, which makes original modification amount deviate from the optimal value. The Finite Element Method was used to verify the deviation of the optimal modification amount under the disturbance of backlash and center distance. Taking the backlash, center distance variation and modification amount as independent variables and the transmission error fluctuation as dependent variables, the mapping relationship between the four factors was established by the Kriging model, Then, the law of the influence of the three variables on the transmission error fluctuation was analyzed, and the optimal modification amount after the backlash and center distance disturbance was obtained. The results showed that when the resistance moment is constant, the increase of modification amount and backlash value leads to the increase of transmission error fluctuation, while the increase of center distance reduces the transmission error fluctuation. Among the three variables, the transmission error fluctuation is the most sensitive to backlash, followed by modification amount and the change of center distance. The modified modification amount reduces the transmission error fluctuation by 2.4% and the contact force fluctuation by 14.9%.

1. Introduction

In the working process of gear transmission system, affected by high temperature environment and various mating clearances, gear backlash and the center distance of gear will deviate from the preset value, which may greatly reduce the effectiveness of the optimal modification based on the original parameter environment in the new parameter environment, and further intensify the meshing vibration of gear [1].

In China, Baolin Wu [2] established the quantitative expression of the relationship between the geometric parameters and the maximum impact force of the meshing impact velocity in the process of gear transmission considering the deformation of the gear under load by applying the theory of mechanical dynamics. Huibo Zhang [3] studied and established the dynamic model of gear rotor considering the coupling of radial clearance and dynamic backlash, and analyzed the backlash of gear radial backlash. On the basis of considering the non-linear characteristics such as backlash, bearing clearance, time-varying meshing stiffness and tooth surface friction. Xiaoan Chen et al [4] put forward a non-linear dynamic analysis method of multi-clearance coupling gear transmission system based on finite element method for the first time. Siyu Chen [5] considered the meshing stiffness and static transmission error of actual gears, studied the dynamic behavior of gears with different modification amount and length. Overseas, Hamed Moradi [6] studied the influence of backlash nonlinearity on the non-linear vibration of spur gears by multi-scale method. Giorgio Bonori [7] estimated the static transfer error by using the nonlinear finite element method and optimized the modification parameters by using genetic algorithm. A Farshidianfar [8] used Melnikov analysis method to study the global homoclinic bifurcation and the transition to chaotic behavior of the nonlinear gear system. However, due to the complexity of the dynamic characteristics of gears under multi-gap coupling and the multi-factors affecting the meshing characteristics of gears, the research in the above literature lacks of the interaction between various factors.

Existing research mainly focuses on the influence of backlash on the dynamic characteristics of gear system under various working conditions, while few studies have been conducted on the influence of backlash and center distance on the optimal modification and the influence of the three on the transmission error fluctuation [9, 10, 11, 12]. In view of the above situation, the following two aspects were studied in this paper: (1) The effects of different backlash and center distance on the transmission error fluctuation and meshing vibration of gears and the offset of the optimal modification amount were compared and analyzed; (2) The influence of the backlash, center distance and the modification amount on the fluctuation of the transmission error was analyzed and the optimal amount of modification after the backlash and center distance disturbance was found.

2. Effect of gear backlash and center distance disturbance on optimal modification

In this paper, a pair of spur gears in a high-speed and light-load aero-gear system is taken as an object to study the influence of gear backlash and center distance disturbance on the optimal modification amount.

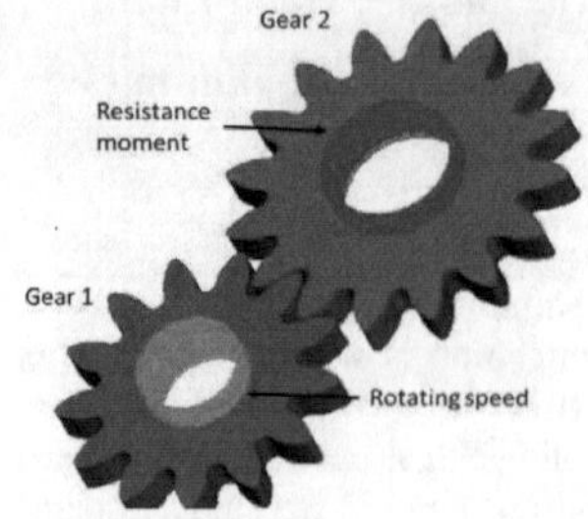

Figure 1. Gear pair of aeronautical gear system

Table 1. Basic parameters of gear pair

Object	Number of teeth	Modulus	Center distance [mm]	pressure angle[°]	Tooth width [mm]	Rounding [mm]
Gear 1	13	1.5	21.75	25	3	0.2
Gear 2	16					

The parameters of the gear pair shown in Figure 1 are shown in Table 1. Gear 1 is the active gear. Speed is 14787r/min. Gear 2 is the driven wheel, and resistance moment is 0.636Nm. The average element size in the contact zone in the gear pair FE model is 0.32 mm. The element types of contact area are hexahedron element and triangular prism element. The contact type in LS-DYNA is "AUTOMATIC_SURFACE_TO_SURFACE". The degrees of freedom in other directions of gears are limited except for the degree of freedom of axial rotation. Since the model does not take into account the influence of shaft and bearing, some simplifications are made in the gear pair FE model:

(1) The influence of elastic deformation of axle end support and bearing clearance is neglected.
(2) The static transmission error caused by manufacturing error and gear accuracy is neglected.
(3) Gear axial movement in meshing is ignored.

Transmission error is the main cause of vibration of gear transmission system [13]. In this paper, the KISSsoft software is used to calculate influence of the parameters including backlash, center distance and modification on transmission error fluctuation of a gear pair transmission. Four different combinations of these three parameters are used for calculation as shown in Table 2 and the corresponding calculations are shown in Figure 2.

Table 2. Model parameters

Number	Modification [μm]	Backlash [mm]	Center distance [mm]
1	20	0.2	21.75
2	20	0.12	21.75
3	20	0.12	21.78
4	15	0.2	21.75

Curves a, b, c and d in Figure 2 correspond to the first, second, third and fourth gear parameters in Table 2, respectively. As can be seen from Figure 2, the fluctuation of transmission error also changes with the change of backlash value and center distance. From the comparison of the transmission error fluctuations of group 1 and group 2, group 1 and group 4, we can see that the increase of backlash and modification amount leads to the increase of transmission error fluctuations, while the comparison of transmission error fluctuations of group 2 and group 3 shows that the transmission error fluctuations decrease with the increase of center distance.

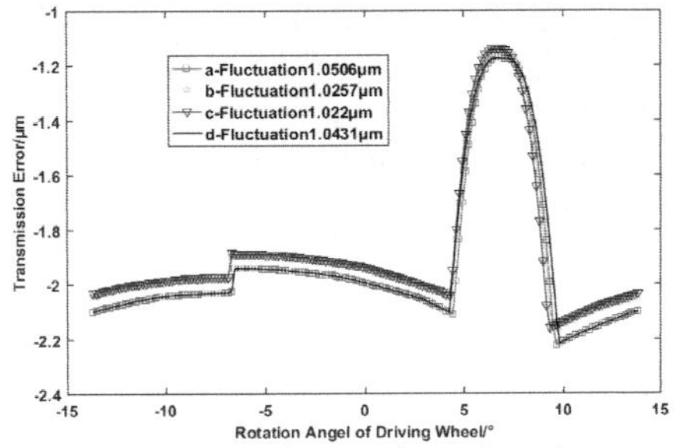

Figure 2. Transmission error fluctuation when the torque is constant

Contact force fluctuation can visually indicate the meshing vibration of gears. Four groups of models in Table 2 are analyzed by ANSYS/LS-DYNA and the contact force of the contact area is obtained as shown in Figure 3.

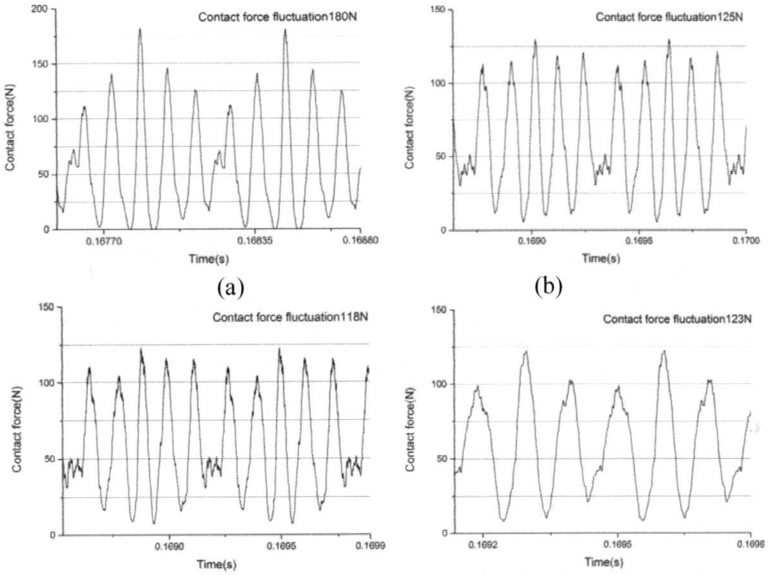

(a) (b)

(c) (d)

Figure 3. Contact force-time curve

In Figure 3, Figures a, b, c and d correspond to the first, second, third and fourth set of gear parameters in Table 2, respectively. The curves in Figure 3 are the two wave periods selected after the contact force fluctuation is stable. Combining with Figure 2 and comparing with Figure b and Figure c in Figure 3, it can be seen that under the same amount of modification, when the backlash value and the center distance change respectively, the contact force fluctuation law and the fluctuation amount of the gear pair will be affected. The increase of the backlash value increases the transmission error fluctuation amount, resulting in the gear pair to increase the fluctuation of contact force by 44%; while the increase of the center distance reduces the fluctuation of transmission error, and the meshing vibration of the gear pair is relieved, but the effect is weaker than the influence of the backlash on the contact force fluctuation. The decline was only 5.6%. Comparing curve a, curve d in Figure 2 and Figure a Figure d in Figure 3, respectively, it can be seen that when the backlash is changed, Correcting the amount of modification can reduce the amount of transmission error fluctuation of the gear pair, Thereby causing the contact force fluctuation of the gear pair to return to the previous level. As shown in Figure 3d, the contact force fluctuation amount becomes 123N, which is only 1.6% different from in Figure 3b. The above analysis shows that the change of backlash and center distance actually makes the original modification amount deviate from the optimal modification amount, which leads to the intensification of meshing vibration of gears. That is to say, the meshing vibration of gear pairs after the disturbance of backlash and center distance can be improved by modifying the modification amount.

3. Determination of the range of parameters

To get the law of influence of backlash, center distance and modification amount on transmission error fluctuation, and then provide direction for correcting amount of modification, an agent model based on Kriging is established.

3.1. Determination of backlash value range

Gear backlash is divided into circumferential backlash j_t and normal backlash j_n. Normal backlash is often used in the calculation of tooth backlash. The formula for calculating the minimum backlash [14] includes two parts, one is necessary for lubrication and the other is necessary due to thermal expansion. The formula for calculating the minimum backlash is as follows.

$$j_{nmin} = j_{na} + j_{nb} \tag{1}$$

$$j_{nmin} = \left(\left|E_{ss1} + E_{ss2}\right|\right) \cdot \cos\beta\cos\alpha - \left|f_\alpha\right| \cdot 2\sin\alpha - K \tag{2}$$

In the formula, $k = \sqrt{\left(fpb1^2 + fpb2^2 + 2(F\beta COS\alpha)^2 + (f_x \sin\alpha)^2 + (f_y \cos\alpha)^2\right)}$

E_{ss1} and E_{ss2} are respectively the upper deviation of tooth thickness of two gears. $f_{pb1} = f_{pb2} = f_{pb}$ are the limit deviation of base section. F_β is tooth tolerance. $F_\beta = F_x$, $f_y = F_\beta / 2f_y$. β is helical angle. α is tooth angle (pressure angle). f_a is limit deviation of center distance.

The backlash value calculated from the above formula combined with the data in Table 1 is 0.28mm. The recommended backlash value of this type of gear in Chinese national standard is 0.08mm to 0.18mm. The recommended backlash value range for this drive system is 0.32mm to 0.4mm. The final determined backlash range is 0.1mm to 0.4mm.

3.2. Determination of the range of center distance variation

At 200 ° C, the distance variation of central pore caused by thermal expansion is the main factor. ANSYS/LS-DYNA was used to analyze the thermal deformation of the box body shown in Figure. 4 at 200 ° C. As shown in Table 3, the initial center distance of gear pair is 21.75mm. The center distance increased by 60μm after thermal expansion. Considering that the center distance may become smaller, it is finally determined that the center distance range is -20 to 60μm.

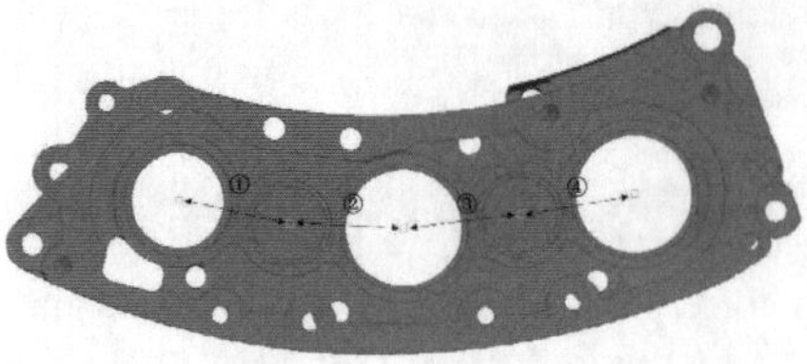

Figure 4. Box model

Table 3. Center distance of box hole

Position	Initial hole spacing[mm]	Hole spacing increment[μm]	Change range of center distance[μm]
1	21.75	60	
2	21.75	60	-20~60
3	21.75	60	
4	21.75	60	

3.3. Determination of the range of modification amount

The profile modification can eliminate the interference of the gear pair due to the deformation of the gear teeth and Smooth gear meshing process. Here, the gear pair is modified in the direction of tooth height. The modified area is shown in the shaded area of Figure 5. The modification parameters mainly include three parameters: the maximum modification amount Δ_{max}, modification length h_{max} and type of modification curve.

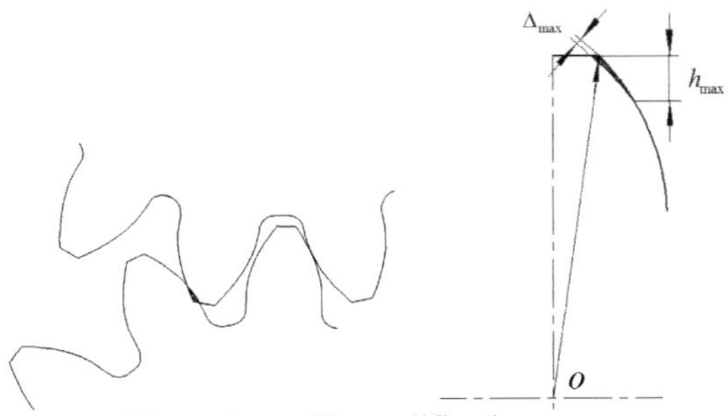

Figure 5. profile modification area

The commonly used methods for estimating the maximum amount of modification are as follows: ideal estimation method, gear manufacturing precision selection method and empirical formula method. Among them, the deformation during the gear meshing process, the gear deformation caused by the temperature and the manufacturing error of the gear is comprehensively considered by ideal estimation method to calculate the maximum shape of the gear. Ideal estimation method is used to calculate the amount of modification here. The formula is as follows:

$$\Delta_{max} = \delta + \delta_\theta + \delta_m \tag{3}$$

In the formula, $\delta = \frac{F_t}{bC_\gamma}$;

$\delta_\theta = \pi \cdot m_n \cdot \cos\alpha \cdot \Delta_\theta \cdot \gamma \times 10^3$;

δ——The amount of loading Deformation of Gear Tooth Surface;

δ_θ——The amount of deformation caused by temperature changes;

δ_m——Gear manufacturing error;

C_γ——Tooth surface comprehensive stiffness;

Δ_{max}——Temperature difference during gear transmission;

γ——Gear material linear expansion coefficient;

Ignore the effects of manufacturing errors in this analysis, so $\Delta_{max} = \delta + \delta_\theta$. Thus, the theoretical tooth profile modification amount after rounding was 10μm. Since the manufacturing error and other factors are neglected, the above theoretical calculation value is on the low side. In order to get a more general rule in a wider range of modification, the modification range is set to 10 to 30μm.

The modified lengths of gear 2 and gear 1 are 0.4899mm and 0.4823mm, respectively.

The modification curve has straight line modification, Walker modification curve, arc modification curve and so on. Arc-modification curve was used here.

In summary, modification parameters are as follows. The amount of modification is 10 to 30μm. The modified lengths of gear 2 and gear 1 are 0.4899mm and 0.4823mm respectively. The modification curve is a circular curve.

4. Agent model construction

4.1. Select sample points

The experimental design theory can help determine the sample points [15]. Common experimental design methods include uniform test design, Latin hypercube design and optimized Latin square design. Here, the optimized Latin square test design is used to select the sample points. In this experiment, 10 sets of samples were selected. The selected sample points are shown in Figure 6.

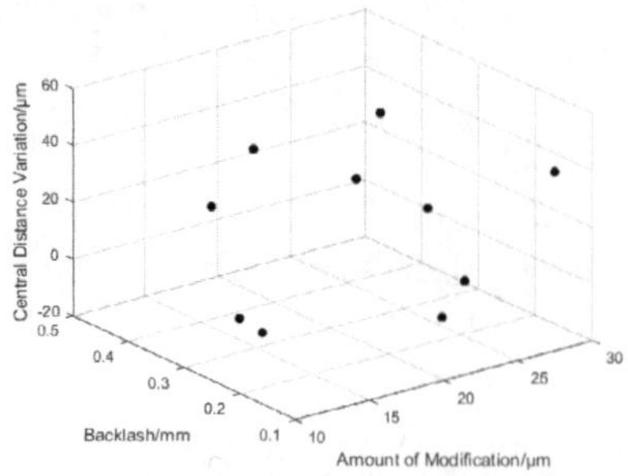

Figure 6. Spatial distribution of sample points

4.2. Kriging model

The Kriging form is as follows:

$$y(x) = \sum_{i=1}^{m} \beta_i f_i(x) + z(x) \tag{4}$$

$f_i(x)$ is the basis function. A polynomial function is generally used. $z(x)$ is the associated stochastic process function. The mean is 0. The process variance σ^2 and spatial covariance functions are defined as follows:

$$\operatorname{cov}(Z(x_i), Z(x_j)) = \sigma^2 R(x_i, x_j) \tag{5}$$

$$\sigma^2 = \frac{1}{p}(y - Xb)^T R^{-1}(y - Xb) \tag{6}$$

In the formula, $R(x_i, x_j)$ is the correlation function of x_i and x_j. The correlation function has different forms, and the Gaussian correlation function is used here:

$$R(x_i, x_j) = \exp(-\sum_{k=1}^{m} \theta_k \left|x_k^i - x_k^j\right|^2) \tag{7}$$

When the correlation function is determined, the response at any point is estimated to be:

$$\hat{y} = \hat{\beta} + \gamma^T(x) R^{-1}(y - f\hat{\beta}) \tag{8}$$

y is a column vector of length n_5, which is the response value of the sample data. When $f(x)$ is a constant, f is a unit column vector of length n_5. $r^T(x)$ is the correlation vector between test point x and sample point $\{x^1, x^2, ..., x^{n5}\}$ which length is n_5.

5. Results analysis

5.1. Model accuracy check

To know the accuracy of the model, 5 sets of data were randomly generated within the sample space. The corresponding values were obtained using the fitted model, and then compared with the calculated values of KISSsoft. As shown in Table 4, A2 indicates the amount of modification. B2 indicates the backlash. C2 indicates the amount of variation in the center distance. y_1 indicates the response value of the fitted model, and y_2 indicates the calculated value of KISSsoft.

Table 4. Comparison between Kriging model response value and KISSsoft calculation result

	Samples				
Number	A2 [μm]	B2 [mm]	C2 [μm]	y_1 [μm]	y_2 [μm]
1	20	0.1	20	1.0514	1.0454
2	25	0.25	60	1.0909	1.0975
3	30	0.325	0	1.1343	1.1417
4	15	0.4	20	1.1402	1.1467
5	10	0.175	40	1.0426	1.0415

Figure 7 shows the values of y_1 and y_2 in a graph form. It can be seen that the relative error is below 1%, which indicates that the Kriging model has high precision and can meet the analysis requirements.

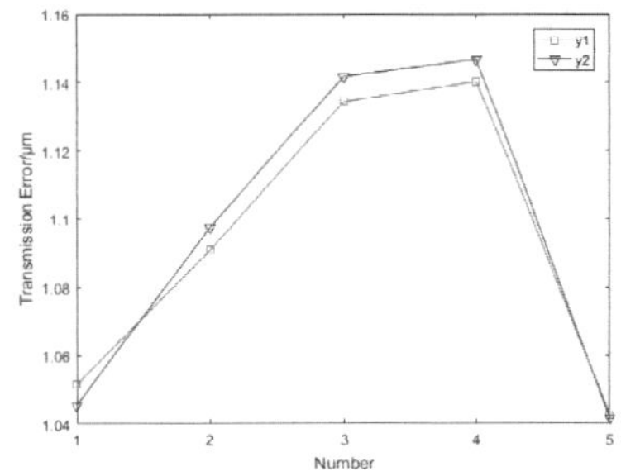

Figure 7. Relative error between Kriging and KISSsoft

5.2. Influence of backlash, modification amount and center distance variation on the fluctuation of transmission error

When the center distance fluctuation, backlash and shape modification are fixed values, the variation of the transmission error fluctuation with the other two variables is shown in Figure 8.

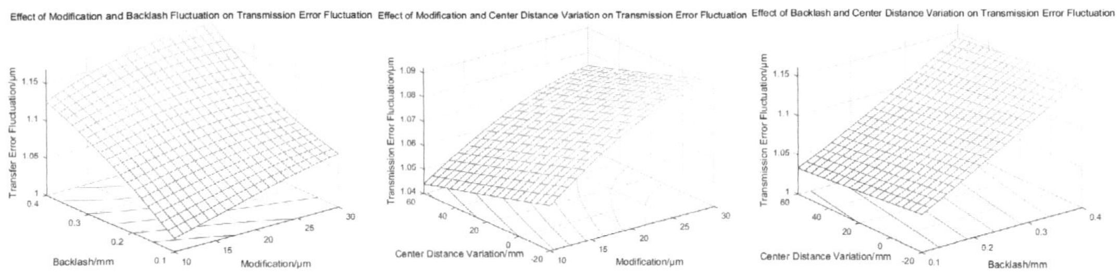

Figure 8. Variation of transmission error fluctuation with variables

It can be seen from Figure 8. that the influence of the amount of modification, backlash and center distance variation on the fluctuation of the transmission error is monotonous when the resistance moment is constant. The fluctuation of the transmission error increases with the amount of modification, increases with the increase of backlash value and decreases with the increase of center distance. In the actual situation, the increase of the center distance will lead to an increase in the backlash. Therefore, the influence of the increase of the center distance on the fluctuation of the transmission error and the meshing vibration has two aspects. (1) The influence of the increase of center distance itself on the fluctuation of transmission error. (2) The influence of the increase of center distance on the fluctuation of transmission error by increasing the backlash value. The sensitivity of transmission error fluctuation to various factors can be obtained by deriving the variables in the fitting model, as shown in Figure 9.

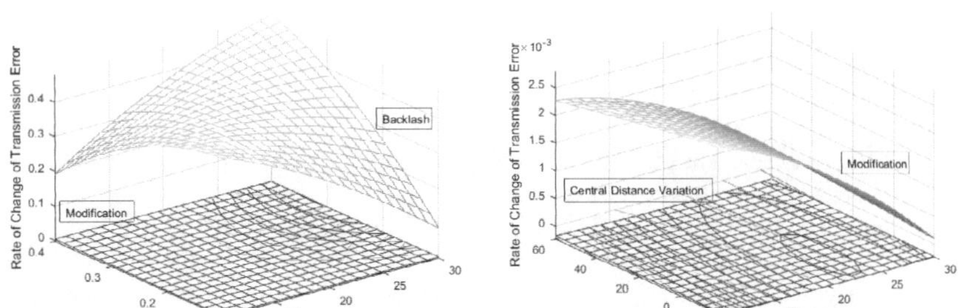

Figure 9. Sensitivity of transmission error fluctuation to variables

It can be obtained from Figure 9. that among the three variables, the transmission error fluctuation is the most sensitive to the backlash, followed by the modification, and the lowest to the center distance variation. From the influence of the variation of the center distance on the fluctuation of the transmission error, the influence of center distance variation on transmission error fluctuation is smaller than that of backlash variation. So in the actual situation, the center distance increases. The positive effect of center distance increase on transmission error fluctuation will be covered by the negative effect of backlash value increase caused by center distance increase. Therefore, the two effects caused by the increase of the center distance are negative after superposition.

5.3. Determination of the optimal modification

When the center distance and the backlash of the gear increase due to thermal expansion, it is known from the influence law of the amount of modification on the fluctuation of the transmission error that the amount of modification should be corrected in the direction of reducing the amount of modification.

Take the gear pair in this article as an example. In actual working conditions, due to thermal expansion, the center distance of the gear pair increases by 60μm to 21.81 mm. At the same time, the backlash value increases by 0.04 mm to 0.36 mm. The original modification amount is 10μm. On the premise that gears do not interfere with meshing and meshing in non-involute area (except modification area), the amount of modification is reduced to 6μm in order to reduce the amount of transmission error fluctuation. On the basis of the above conditions, the calculation of the four sets of models in Table 5 is performed to check whether the corrected amount of the modification is valid. The amount of transmission error fluctuation and the contact force fluctuation are shown in Figure 10 and Figure 11, respectively.

Table 5. Simulation model

Number	Backlash [mm]	Modification [μm]	Center distance [mm]
1	0.32	10	21.75
2	0.36	10	21.81
3	0.36	6	21.81
4	0.36	0	21.81

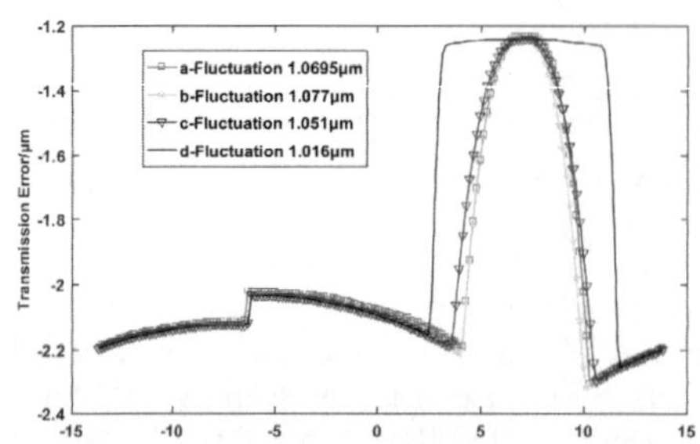

Figure 10. Fluctuation of transmission error

The first, second, third, and fourth sets of models in Table 5 correspond to the curves a, b, c, and d in Fig 10 and Figures a, b, c and d in Figure 11 respectively. Referring to Table 5 and Figure 10 and Figure 11, the comparison of the first and second sets of models shows that when the backlash and the center distance increase, the transmission error fluctuation and the contact force fluctuation of the gear pair increase. The increase of transmission error fluctuation and contact force fluctuation is 0.7% and 15.2% respectively. When the modification amount is reduced to 6μm, it can be seen that the transmission error fluctuation and the contact force fluctuation amount return to previous levels. The transmission error fluctuation was reduced by 2.4% and the contact force fluctuation was reduced by 14.9%. Continue to reduce the amount of modification to 0. That is the fourth group of models. The transmission error fluctuation amount is further reduced, but the contact force fluctuation amount of the gear pair increases and the contact force fluctuation amount increased up to 22.7% compared with the result of the modification of 6μm. The reason for that is that the too small amount of modification causes the gear pair to interfere or engage in non-involute zone engagement. When the amount of modification can ensure that the gear pair does not interfere and the gear pair does not engage in non-involute zone (except for the modification zone), the amount of modification is the optimal amount of modification. After the thermal expansion causes the backlash and the center distance to increase, the transmission error fluctuation increases, which also causes the optimal modification amount to shift to a smaller value. Therefore, the amount of modification should be reduced so that the amount of modification is optimum in a new parameter environment.

In summary, in the previous theoretical calculation method of shape modification, the influence of backlash and center distance on the amount of shape modification is often neglected. It can be seen from the above analysis that the disturbance of backlash and center distance will make the original modification amount Deviate from the optimal value. And by correcting the amount of modification, the meshing vibration of the gears pair can be reduced to some extent after backlash and center distance disturbance. It can be further understood that there are different amounts of optimum modification under different backlashes and center distances.

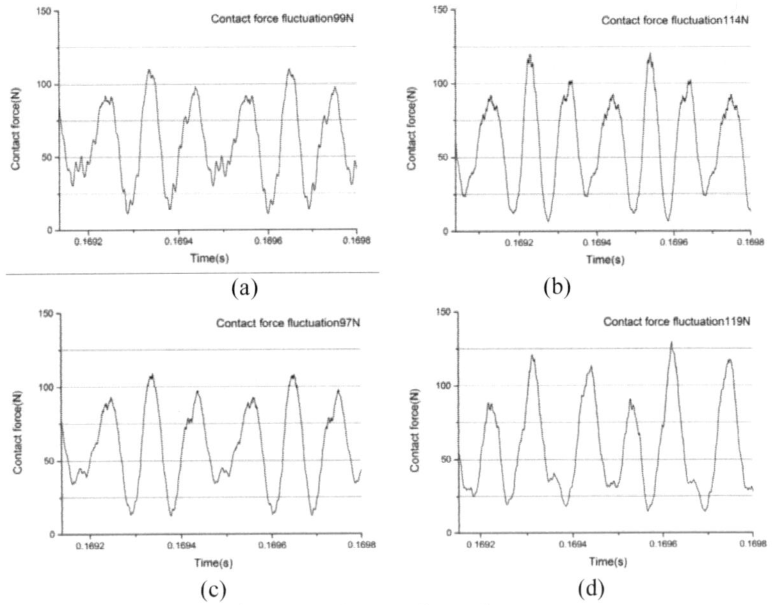

Figure 11. Contact force-time curve

6. Conclusion

(1) The disturbance of backlash and center distance increase the fluctuation of the transmission error, which in turn exacerbates the meshing vibration and increases the fluctuation of contact force.

(2) It was verified that the original modification amount is deviated from the optimal value when the backlash and center distance disturbance.

(3) The Kriging-based agent model was established and its accuracy was verified. Based on that, the influence of the modification amount, backlash and center distance variation on the fluctuation of the transmission error was analyzed. The results showed that without considering the fluctuation of torque, the influence of the amount of modification, backlash and center distance variation on the fluctuation of the transmission error was monotonous. The increase of the amount of modification and the value of the backlash will increase the fluctuation of the transmission error, and increasing the center distance helps to reduce the fluctuation of transmission error. In the three variables, the transmission error fluctuation is the most sensitive to the backlash. The modification amount is the second and the sensitivity to the center distance variation is the lowest. It provides a reference for understanding the influence of modification, backlash and center distance variation on transmission error fluctuation and meshing vibration.

(4) The optimal modification amount of the gear after the backlash and center distance disturbance was found. The results showed that the modified amount of modification reduced the transmission error fluctuation by 2.4% and the contact force fluctuation by 14.9 %. It provides an idea for correcting the amount of modification after backlash and center distance disturbance.

7. References

[1] G H LI, Y T FEI. Influence of temperature change on the meshed gear` s backlash [J]. Journal of Hefei University of Technology (Natural Science), 2004, 27(10):1147-1150.

[2] B L WU, S J YNAG, J H YAO. Theoretical Analysis on Meshing Impact of Involute Gears [J]. Mechanical Science and Technology, 2003, 22(1):55-57.

[3] H B ZHANG, R WANG, Z K CHEN, et al. Nonlinear dynamic analysis of a gear-rotor system with coupled multi-clearance [J]. Journal of Vibration and Shock, 2015, 34(8):144-150.

[4] X A CHEN, Y Y MIU, W YANG, et al. Nonlinear Dynamic Characteristics Analysis of Multi-Gap Coupling Gear Transmission System Based on Finite Element Method [J]. Journal of Vibration and Shock, 2010, 29(2):46-49.

[5] S Y CHEN, J Y TANG, Z W WANG, et al. Effect of Modification on Dynamic Characteristics of Gear Transmissions System [J]. Journal of Mechanical Engineering, 2014, 50(13):59-65.
[6] Moradi H, Salarieh H. Analysis of nonlinear oscillations in spur gear pairs with approximated modelling of backlash nonlinearity [J]. Mechanism & Machine Theory, 2012, 51(5):14-31.
[7] Bonori G, Barbieri M, Pellicano F. Optimum profile modifications of spur gears by means of genetic algorithms [J]. Journal of Sound & Vibration, 2008, 313(3):603-616.
[8] Farshidianfar A, Saghafi A. Global bifurcation and chaos analysis in nonlinear vibration of spur gear systems [J]. Nonlinear Dynamics, 2014, 75(4):783-806.
[9] Y M LI, W D CEHN. Q CHEN, et al. Research of the Influence of Gear Backlash on Gear Pair Nonlinear Vibration Characteristic [J]. Journal of Mechanical Transmission, 2013(5).
[10] X P LI, J X MOU, W J PAN, et al. Influence of Fractal Backlash on Dynamic Behavior of Gear-bearing System [J]. Journal of Mechanical Transmission, 2018, 54(9):153-160.
[11] C LI, X P XI, H Y WU, et al. Analysis on Tooth Backlash to Dynamic Influence of Meshing Force [J]. Missiles and Space Vehicles, 2018(3).
[12] H H HE. Effect of Backlash on Dynamics Behavior of Gear System [J]. Mechanical Design and Manufacturing, 2012(7):281-283.
[13] J WANG, J ZHANG. Effects of gear modification on transmission error of a helical gear pair [J]. Journal of Vibration and Shock, 2018, 37(2):254-260.
[14] S HU. Gear Side Clearance Calculation Method [J]. Metal Working, 2008(4):28-29.
[15] T Z LI, G Y LI, T CHEN, et al. Robustness Design of Occupant Restraint System Based on Kriging Model [J]. Journal of Mechanical Transmission, 2010, 46(22):123-129.

IOP Conf. Series: Materials Science and Engineering **638** (2019) 012017 doi:10.1088/1757-899X/638/1/012017

Study on the compacting device of combination for ridge-till and no-till planters

Hongqi Luo[1] , Xia Liu[2*], Jiatong Qin[1] and Guangwei Zhao[1]

1 School of Material and Mechanical Engineering, Beijing Technology and Business University, Beijing, China
2 Quality Management Branch of China National Institute of standardization, Beijing, China
E-mail: lhqfrq@126.com
*Corresponding author

Abstract. Ridge tillage and conservation tillage combination is an important technology. The compacting device is one of the key components of no till planter. According to different agronomic requirements, a general type of combined type of pressing device was designed. In the device, the compacted wheel set should be designed, the ridge is compacted along the loose soil, and the ridging effect is enhanced. Meanwhile, the ring structure is designed on the compacting wheel to suppress the seeding belt. Through the force analysis and Simulation of the suppression device, the structural parameters are determined, and the adjustable mechanism is designed to realize the adjustability of the pressure of the town. By designing movable pressing ring, the number and size of the ring can be adjusted to meet the needs of different agronomic conditions. The modelling of combined press device is conducive to the promotion of no tillage seeding technique, which is a better combination of advantages of ridge tillage and conservation tillage.

1. Introduction

Because ridge-till and no-till planter is operating in the original ridge tillage with crop stubble and weed cover, it will lead to a large number of soil blocks, coupled with the impact of stubble weeds, so that the seeds cannot be very good contact with the surrounding soil. It is more difficult to sow and suppress than traditional tillage method [1, 2]. Therefore, according to the physical properties of soil and the form of ditch opener in no-tillage operation, it should select or design a good and reliable soil covering and suppressing device.

The traditional development process of agricultural machinery product has to go through many design steps, such as prototype design, trial production, field test, improvement design, re-trial production, re-test and so on, resulting in high cost and long cycle. With the application of information technology, computer technology and automation technology in machinery manufacturing industry, new vitality has been brought to agricultural machinery manufacturing industry. Using virtual prototyping technology to develop agricultural machinery products, designers can build prototype models on computers, carry out structural analysis and dynamic analysis of the prototype according to research purposes, and replace traditional physical experiments with digital simulation [3, 4]. Because the virtual design of components is a three-dimensional entity design, and can endow all kinds of real materials, which can obtain parameters information such as volume, quality, centroid position, cross-section area, cross-section inertia moment, and cross-section bending and torsion modulus directly from the computer [5]. It can provide convenience for designers to make decisions and administrators to approve the cost, thus realizing the whole product. Optimum design of performance before actual manufacture. In addition, the virtual prototyping technology is used to evaluate the product assembly in the product development stage, so that the manufacturability of the product can be considered from the assembly point of view of the whole product in the design stage, rather than from the point of view of a single part, avoiding design errors, providing convenience for future production finalization and saving time. In addition, the traditional two-dimensional drawings design cannot find the assembly interference caused by the design in advance, and the results often wait until all parts are processed and manufactured before they are actually assembled, resulting in a waste of time and money. By using assembly process simulation software and virtual assembly technology, designers can inspect the assembly clearance and interference between parts in an all-round way before the prototype is actually processed.

2. Agronomic Situation and Design Ideas of the Combination Press

The single-ridge and multi-row planting mode is a new technology for planting multi-row crops on ridges. The combination of large-ridge planting and conservation tillage technology is a new technology. Especially when the ridge spacing matches the wheel spacing of tractor, the machine works in the field. All the wheels are running in fixed ridges and furrows. Because there is no machine to suppress the crop growth belt, it does not need to plough or loosen deeply every year, so it can keep good crop life. In no-tillage mode with large ridges, wheels run in fixed ridges and furrows. The width of ridges matches the distance of tractor wheels, and the tractor is guided by ridges and furrows, which is helpful for furrow openers to avoid residual corn stubble and realize seeding without corn stubble [6]. The planter equipped with the combined suppressor seeded six rows of wheat on the ridge of two rows of corn stubble, aiming at the agronomic requirements of two cropping areas and high stubble of Maize in one year.

Because the planter requires no-tillage planting on the big ridge, the original ridge will be damaged by man-made, machine operation and wind and rain. In order to realize no-tillage planting on the original ridge and ensure the quality of planting, it is necessary to repair the damaged area of the big ridge. However, after Ridge restoration, the soil along the ridge is loose and easy to collapse, so it is necessary to repair it. Compaction of loose soil along the ridge after restoration can solve the problem that the soil along the ridge is easy to fall into the ridge and furrow after restoration. Therefore, it is necessary to design a suitable combined compactor to suppress both the ridge and the planting belt at the same time.

Force Analysis and Parameter Determination of Pressure Wheel

The forces of no-tillage machine when planting include self-gravity, spring pressure, and tractor pulling force acting on monomer through frame, working resistance of ditch opener and reaction force of soil to crushing wheel. The supporting force of the single articulation point and the rolling resistance of the roller bearing are neglected [1]. The whole force analysis of the planter is shown in Fig. 1.

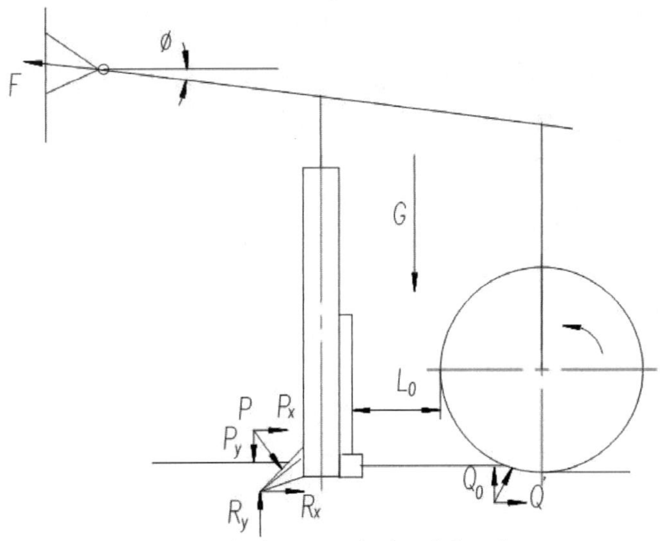

Fig. 1 Whole force analysis of the planter

Where:

G - Single body weight (including spring pressure)

R_x - Horizontal working resistance of ditcher

R_y - Vertical working resistance of ditcher

Q_0 - Reaction force of soil to the wheel

Q' - Rolling resistance of pressure wheel

P - Soil reaction force

F - Monomer tension

φ - The angle between tension and horizontal line (clockwise direction is positive)($\varphi = 15°$)

If these forces are decomposed along the Y-axis perpendicular to the tie rod, then

$$\sum Y = 0$$

$$Q_0 \cos\phi + R_x \cos\phi - G\cos\phi + Q' \sin\phi + P_x \sin\phi - P_y \cos\phi = 0 \tag{1}$$

Where: $Q' = Q_0 f_0$

f0 - Rolling Resistance Coefficient (f0=0.28)

According to the principle that the action force and reaction force are equal, the action force of the crushing wheel on the ground in the vertical direction is Q= Q_0, then put Q′ into the Eq. 1. It can obtain Eq.2.

$$Q = Q_0 = \frac{G - R_y - Rxtg\varphi - P_x tg\varphi + Py}{1 + f_0 tg\varphi} \tag{2}$$

The working resistance Rx and Ry are related with the shape of the opener, the work speed of the opener and the seeding, which can be expressed in the following Eq. 3 [1].

$$R_X = A_X + B_X V + C_X h_2$$
$$R_y = A_y + B_y V + C_y h_2 \tag{3}$$

Where: V- Planting speed (km/h); H- Ditching depth (cm)
Ax, Bx, Cx, Ay, By, Cy - The regression coefficients related to the type of ditcher. For sharp-angle ditcher, Ax = 7.3701, Bx = 1.2658, Cx = 0.3626, Ay = 4.4215, By = 0.3512, Cy = 0.0893.
If Planting speed V is 5km/h and furrow depth h is 5cm, which is calculated by substitution formula: Rx = 22.7641N, Ry = 8.41N, driving traction F = 700N, vertical traction N = 55N, so Px = F - Rx = 677.2359N, Py = N - Ry = 46.59 N. If the weight of no-tillage planter is 300 kg, and plus weight and spring pressure is 100 kg, then G is 4000 N. By substituting Rx, Ry, Px, Py , f0, it can get Q = 2396 N, then Q′ = 670 N.
The diameter of the crushing wheel has a great influence on the crushing effect. Generally, when the wheel weight is equal, the smaller the wheel diameter, the greater the soil breaking effect and pulling force. The force analysis is shown in Fig. 2. According to the principle of moment balance,

$$\sum A = 0$$
$$H \bullet r = R_y \bullet s + R_x \bullet h$$
$$\sum C = 0$$
$$H \bullet (\mathrm{r} - h) = G \bullet S$$

By combining two formulas, it can get the Eq. 4.

$$H \bullet \mathrm{r} = G \bullet s + H \bullet h = R_y \bullet s + R_x \bullet h \tag{4}$$

In the formula: H-traction, G-Wheel Weight and Spring Pressure, R-wheel radius, S-wheels advancing when sinking 2.64 cm, R-Supporting Force, V-Speed of Wheel Forward, T-the time when the wheel advances.
When the traction force and wheel weight are fixed, the wheel diameter is smaller and the compaction time is shorter, the compaction effect of soil will be less ideal. The slip coefficient will increase, which will easily drag the soil back. The results show that the diameter of the wheel should be 20-50 cm, and the diameter of the roller is selected as 40 cm.

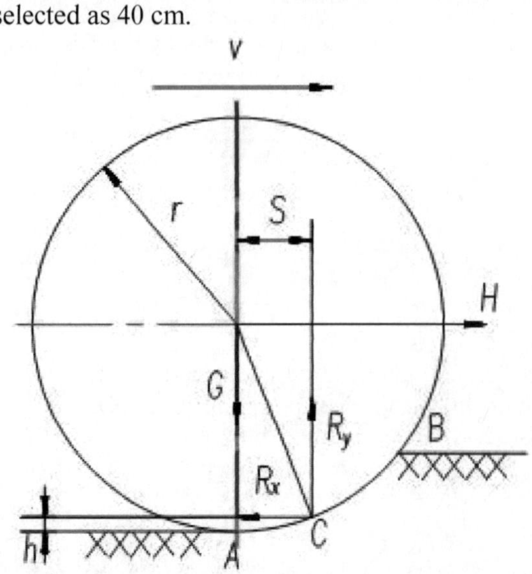

Fig. 2 Stress analysis of the compacting roller

3. Modeling of Composite Compression Device

The combined crusher mainly includes compacting wheel, crushing wheel axle, wheat crushing wheel I, wheat crushing wheel II, spacer tube and other parts. The compacting wheel is welded by the hemisphere of the compacting wheel and the welding box of the end plate of the compacting wheel [7,

8]. The model of the parts is shown in Fig. 3. The model of the wheat crushing wheel I and the crushing wheel II is shown in Fig. 4.

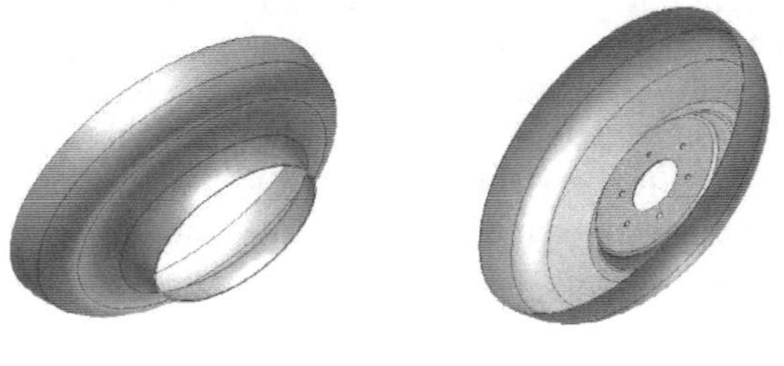

Fig.3 Compacting wheel

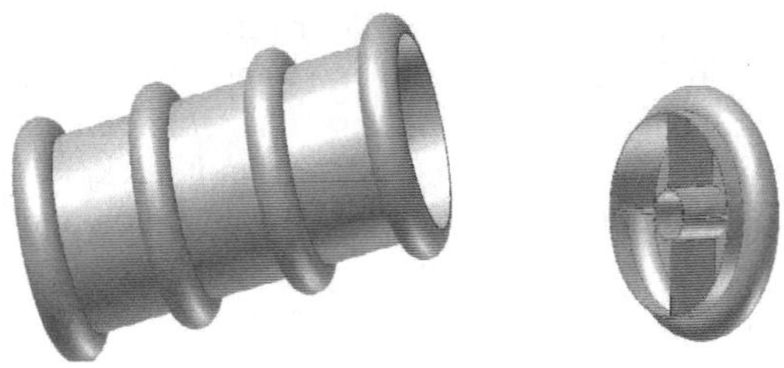

Fig.4 Crushing wheel

Fig. 5 Assembly of combined compacting device

After the parts model of the No-tillage Planter is drawn, it enters the assembly environment. In the assembly drawing of the combined compressor, the compacting wheel, the wheat crushing wheel I, the wheat crushing wheel II and the spacer pipe are connected with the compressing wheel axle through insertion constraints and motion constraints, while the rod-pulling device of the compressing wheel is connected with the compressing wheel axle through the wear-resistant axle, mainly using insertion and motion constraints to assemble. The overall assembly results are shown in Fig. 5. The compaction wheel is used to compact the loose soil along the ridge after Ridge pruning to enhance the effect of ridge pruning, and the planting belt relies on the ring structure on the roller I and the roller II to achieve the suppression. The spacing canal is the corn stubble zone. Because the common stubble is strong in the protuberance of about 50 mm on the surface, it is easy to lift the compressor, which affects the suppression effect. Therefore, the spacing canal is used to open the gap in the stubble zone, and the distance between the spacing canal and the surface is about 70 mm during planting time, which weakens the uplifting effect of the corn stubble [8].

Due to different agronomic conditions and tillage systems, the demand of no-tillage planters in different areas is different (such as ridge height, width, ridge shape, row spacing, plant spacing, etc.). Through the establishment of the universal parts design library of no-tillage planter, the size of parts can be changed by modifying the parameters to meet different needs.

4. Conclusion

A general combined crushing device was designed for different regions with obvious agronomic differences. The device belongs to a combined type of suppression, and simultaneously suppresses the seeding belt and ridge edge. A compaction wheel was designed to compact the loose soil along the ridge after Ridge pruning to enhance the effect of ridge pruning and solve the problem that the soil along the ridge is easy to fall into the ridge and furrow after Ridge pruning. The device is universal. Its parameters are determined by the force analysis of the roller. The roller shaft is connected with the connecting rod of the roller through the wear-resistant shaft. The roller tie rod can adjust the rotary shaft by adjusting the nuts and the pressure, and the magnitude of the pressure can be adjusted according to the need. An adjustable crushing ring is designed, which can be refitted according to the actual needs of the number and thickness of the ring to meet the needs of different agronomic conditions. The three-dimensional model of the combined crushing device is established, which is helpful for structural analysis and dynamic analysis according to research purposes, and for checking assembly clearance and interference between components.

5. Acknowledgments

Project is supported by National Key R & D Program of China (2017YFF0207200), Beijing Organizational Department of Excellent Talents Training Subsidy Project (19000530803) and 2017 Beijing University Students Research and Entrepreneurship Action.

6. References

[1] China Academy of Agricultural Mechanization Science. Handbook of Agricultural Machinery Design (Volume 1). Beijing: Machinery Industry Press. 1988.

[2] He Jin, Li Hongwen, ChenHaitao, et al. Research progress of conservation tillage technology and machine[J] . Transactions of the Chinese Society for Agricultural Machinery, 2018，49 (4):1-19.

[3] Yang Minqin. Mechanics of Materials [M]. Beijing: Machinery Industry Press, 1994.

[4] Su Yuansheng. Research on No-tillage Ditcher and Its Testing System [D]. Beijing: China Agricultural University, 1999.

[5] Jia Honglei, Zheng Jiaxin, Yuan Hongfang, et al. Design and experiment of profiling sliding-knife opener [J] . Transactions of the CSAE, 2017，33 (4):16-24.

[6] Jia Mingyu. Experimental study and computer aided design of no-tillage planter's suppression device [D]. Beijing: China Agricultural University, 2000.

[7] Gao Yulu. Study on the Slipping Phenomenon of Land Wheel of No-tillage Planter [D]. Beijing: China Agricultural University, 2001.

[8] Luohongqi. Research on ridge-tillage No-tillage Planter for corn stubble field [D]. Beijing: China Agricultural University, 2006.

IOP Conf. Series: Materials Science and Engineering **638** (2019) 012018 doi:10.1088/1757-899X/638/1/012018

Effect of Ru on the concentration distribution of elements and creep properties of nickel-based superalloy

Shuang Liang, Zhixin Liu

Department of Mechanical and Power Engineering, Yingkou Institute of Technology, Yingkou 115000, china
E-mail: 70080952@qq.com

Abstract. The effects of Ru on creep properties, microstructure and elemental concentration distribution of nickel-based single crystal superalloy containing without Ru and 2% Ru are studied by means of creep tests under high temperature, morphology observation after heat treatment and determination of elemental concentration distribution by atomic probe. The results show that Ru can inhibit the precipitation of TCP phase and improve the solid solubility of W element in the alloy and the creep life of the alloy at high temperature. At the same time, the addition of element Ru can make the alloy to produce "reverse distribution" effect, improve the alloying degree of γ' phase and the effect of solution strength, which final significantly improve the high temperature creep rupture properties of the alloy containing Ru.

1. Introduction

Nickel-based single crystal alloy is a kind of superalloy with nickel as the matrix, which is composed of Al, Co, Cr, W and other elements. It is prepared by directional solidification technology, which has been used in aerospace, warships, power generation, chemical industry and other important industrial fields due to good high temperature strength, oxidation resistance, corrosion resistance and fatigue resistance [1,2].

The addition of Ru is one of the important signs of the fourth generation of nickel-based single crystal superalloy. It was found that Ru is slightly segregated in the dendrite regions during solidification, and Ru had not obviously effect on the segregation behavior of other elements in the inter-dendrite/dendrite regions [3,4]. Other studies showed that no obvious the Ru is segregated in the inter-dendrite/dendrite regions [4,5], but can reduce the segregation of W, Mo and other refractory elements. The morphology of the alloy containing Ru showed that Ru can reduce the size of γ' phase and increase the cubic degree of γ' phase, which it can make the mismatch degree of γ /γ' two phases more negative[6,7]. However, it has been reported that Ru has no significant effect on the mismatch of γ /γ' two phases [8]. O'Hara et al.[9]found that Ru can change the distribution ratio of elements in the two phases of γ /γ', and propose the "reverse distribution" effect. The so-called "reverse distribution" effect refers to the diffusion of element that was mainly segregated in γ' phase to γ phase by adding Ru, thus reducing the concentration of the element in γ' phase and increasing the concentration in γ phase.

The element which was mainly segregated in γ phase diffused into the γ′ phase, which reduced the concentration of the element in the gamma phase and increased the concentration in the gamma phase.

Ofori et al.[10]showed that Ru can indeed change the distribution ratio of elements in two phases of γ /γ′, especially Ru can significantly reduce the distribution ratio of Re, at the same time, the addition of Ru can reduce the possibility of TCP phase precipitation. However, some studies have different conclusions, for example, Cui et al. believed that Ru had no effect on the distribution behavior of elements Cr, Mo and Co, and that Ru should not inhibit the precipitation of TCP phase.

In this paper, 2%Ru alloy was designed and prepared. By means of observation of microstructure, determination of element concentration and creep properties test, the effects of Ru on element concentration distribution, high temperature creep properties and morphology of nickel-based single crystal alloys were studied, and which is tried to find a way to prepare high performance Re-free nickel-based single crystal alloys containing Ru for providing theoretical basis and data support in future alloy composition design.

2. Experimental materials and methods

Three nickel-based single crystal alloys have been designed and prepared, which their mass fraction and atomic fraction are shown in Table 1. The main difference between the three alloys lies in the presence or absence of Ru elements. Ru-free alloys containing 4%W is defined as alloy 1 and Ru-free alloys containing 6%W as alloy 2, and 2%Ru alloys containing 6%W is defined as alloy 3.

Table 1. True chemical composition of the alloys.

	Al		Ta		Cr		Co		Mo		W		Ru	
	wt.%	at.%	wt.%	at.%	wt%	at.%	w%	at.%	wt.%	at.%	wt.%	at.%	wt%	at.%
Alloy 1	5.98	13.35	7.62	2.54	5.82	6.74	5.97	6.10	6.25	3.92	4.07	1.33	0	0
Alloy 2	6.00	13.46	7.65	2.55	5.80	6.58	5.90	6.12	6.25	3.92	5.99	2.04	0	0
Alloy 3	5.99	13.69	7.55	2.53	5.80	6.58	5.89	6.12	6.07	3.81	5.96	2.02	1.98	1.20

DSC curves of the three alloys show that the initial melting temperatures are about 1323°C. Thereby, the heat treatment process of the alloy is determined as follows: 1280°C × 2 h, A. C + 1315°C × 4 h, A. C + 1070°C × 4 h, A. C + 870°C × 24 h, A. C (A.C is air cooling). After completely heat treatment, the sheet creep specimens with cross section of 4.5mm x 2.5mm and gauge length of 20mm were prepared. After mechanical grinding and polishing, the creep samples were putted into the creep testing machine named GTW504 to test the creep properties under different conditions.

After mechanical grinding and polishing of alloys with different compositions and states, chemical corrosion was carried out. The composition of corrosion solution was HCl (38%, 100ml) +$CuSO_4 \cdot 5H_2O$ (20g) + H_2O (80ml) + H_2SO_4 (98%, 5ml). The corroded specimens were cleaned by ultrasonic and observed by SEM. The element concentration distribution of γ/γ′ two-phase in Ru-free alloy and 2%Ru alloy under different states were measured by atomic probe. The needle-like sample of 0.3 mm × 0.3 mm × 14 mm were prepared along the orientation of [001], and that were inserted into the three-dimensional atomic probe (3DAP) of LEAP4000XHR, which can determination element concentration distribution after electro polishing.

3. Results and Analysis

3.1 Effect of W on microstructure

After complete heat treatment, the microstructures of three alloys with different compositions are observed as shown in Figure 1, in which Figure 1(a) is the morphology of free-Ru/4% W alloy, Figure 1(b) is the morphology of free-Ru/6% W alloy and Figure 1(c) is the morphology of 2% Ru/6% W alloy.

It can be seen from the Figure 1 that the structure of three alloys is basically the same, all cubic γ' phases are embedded in the matrix of γ phase with a coherent manner, and the cubic γ' phases are arranged regularly in the direction of <100>. Among them, the size of γ' phase in alloy 1 and alloy 2 is about 0.46μm, the size of matrix channel is about 0.1μm, and the volume fraction of γ' phase is about 62%. The size of γ' phase in alloy 3 is about 0.42μm, the size of matrix channel is about 0.05μm, and the volume fraction of γ' phase is about 67%. TCP phase taken on short rod can be seen in alloy 2 as shown in the white arrow of Figure 1 (b). It is concluded that the difference of three alloys lies in the different contents of Ru and W. Compared with alloy 1, the W content of alloy 2 increased can led to. Compared with alloy 2, alloy 3 adding 2% Ru has not precipitation of TCP phase. Therefore, it can be inferred that the increase of W content is the main cause of precipitation of TCP phase, while the addition of Ru inhibits the precipitation of TCP phase.

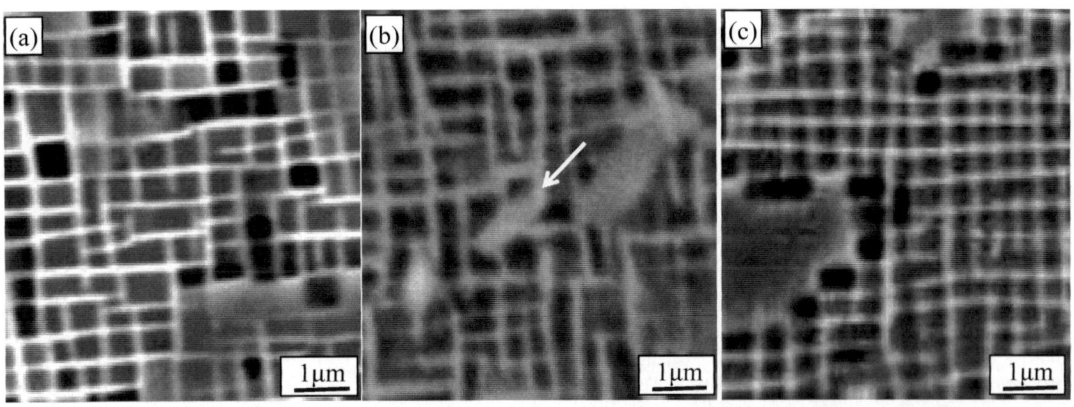

Figure 1. Morphology of (a) free-Ru/4%W alloy, (b) free-Ru/6%W alloy and (c) 2%Ru/6%W alloy after full heat treatment.

3.2 Creep Property of alloy

The creep properties of three alloys after fully heat treatment were tested under high temperature. The experimental conditions were 1100°C/137MPa. The creep curves were drawn as shown in Figure 2, in which curves 1, 2 and 3 represented alloy 1, alloy 2 and alloy 3 respectively. It can be illustrated that the creep characteristics of the three alloys are obviously different. The creep life of alloy 3 is the longest of 182 hours, and creep can be divided distinctly into three stages, that is, initial stage, steady stage and acceleration stage, in which the steady stage lasts the longest time of 125 hours, and the strain rate is smaller of 0.0063%/h. Compared with alloy 3, alloy 2 has a longer creep life of 98h, and the three creep stages are not obvious, especially the strain rate value is bigger of 0.0527%/h in steady creep stage. The creep life of alloy 1 is the shortest of 42h, and the creep curve tends to be linear, and which is impossible to distinguish three stages of creep by creep curve.

The results show that the three alloys have different creep life. It can be inferred that the presence/absence of Ru and the different content of W are the main factors affecting the creep properties of the alloys under high temperature. It is considered that W of refractory element is the main solid solution strengthening element in nickel-based single crystal alloy, which the increase of W content can improve the high temperature properties, but W is also one of the main forming elements of TCP phase. The increase of W content can promote the precipitation of TCP phase and lead to the rigidly decrease of creep properties. The addition of Ru inhibits the precipitation of TCP phase. Therefore, the interaction between Ru and W is the main reason for the significant improvement of creep properties of the alloy.

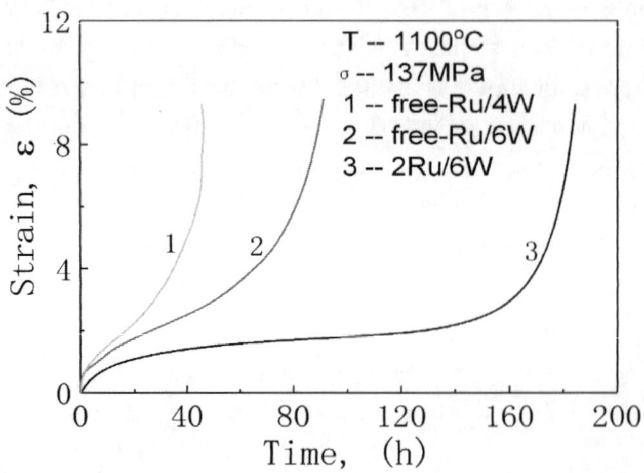

Figure 2. Creep curves of different alloys at 1100°C/137MPa.

3.3 Effect of Ru on element concentration distribution

Three-dimensional atomic probe (3DAP) was used to determine the concentration distribution of each element in the γ/γ′ two phases of the two alloys. The concentration distribution of elements Ta and Mo was used to explain the test area and the distribution of element concentration, as shown in Figure 3. Figure 3 (a) and Figure 3 (b) are the concentration distributions of Ta and Mo in Ru-free alloys, which the dark region in the figure represents the higher concentration of the element and the light region represents the lower concentration of the element. According to the concentration distribution characteristics of Ta and Mo in single crystal nickel-based alloys, that is, Ta mainly distributes in γ′ phase and Mo mainly distributes in γ phase, and thus it can be judged that area A is γ′ phase and area B is γ′ phase. Figure 3 (c) and Figure 3 (d) are the concentration distributions of Ta and Mo in 2%Ru alloys. Using the same method to judge the γ/γ′ two-phase region, and it can be seen that the A region is γ′ phase and the B region is γ phase. Take Figure 3 (c) as an example to illustrate the measure selection of concentration of atomic, and cubes of 25 nm x 30 nm are selected as the measurement areas of phase concentration in different phase regions. As shown by the square in the figure, only one phase area is shown in the figure, while the other phase area is the same, but not shown in the figure. In order to measure the concentration gradient at the interface of two phases, a cylinder measuring area of 10nm x 50nm is taken at the interface of two phases, as shown by the cylinder in the figure.

The average concentration of elements in the γ/γ' phase in the Ru-free and 2%Ru alloy is determined as showed in Table 2. In order to compare the concentration distribution of elements in the two phases, the distribution ratio is introduced, that is, the average concentration of an element in the γ' phase divided by the average concentration of the element in the γ phase. The smaller concentration value is defined as 1, whereas the relatively larger part is defined as the multiple of 1. From Table 2, it can be seen that the distribution trend of elements in γ/γ' phase is the same, in which Al and Ta as the forming elements of γ' phase are mainly distributed in γ' phase, while Mo and W are mainly distributed in γ phase, and Ru is mainly distributed in γ phase. However, the distribution ratios of elements in γ/γ' phases are significantly different, and the distribution ratios of Al elements in alloy 2 are significantly lower than alloy 1, while the distribution ratios of W and Mo elements are significantly higher. This is attributed to the "reverse distribution" effect, that is, the diffusion of elements mainly segregated in γ' phase to the γ phase results in the decrease of the concentration of the element in the γ' phase and the concentration in γ phase. The concentration of the element mainly segregated in γ phase diffused to the γ' phase, which can reduce the concentration of the element in the γ phase and increase the concentration in γ' phase.

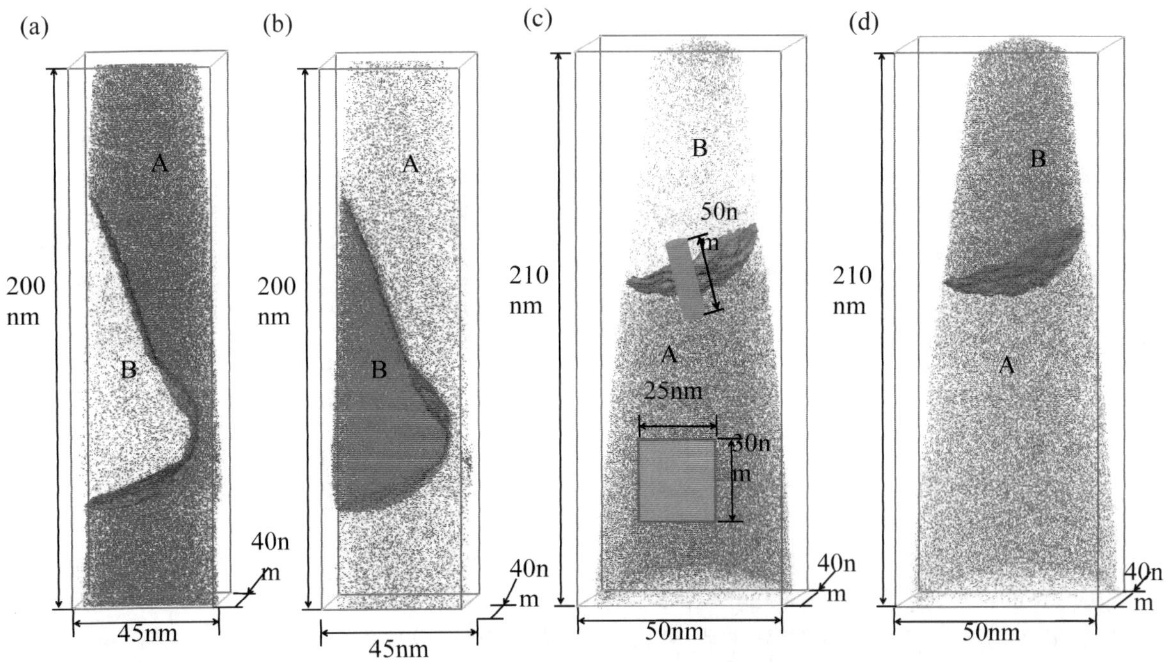

Figure 3. Schematic maps of Ta and Mo concentration distributions and the ROI positions in (a) Ta—Ru-free alloy, (b) Ru-free alloy, (c) Ta-2%Ru alloy and (d) Mo-2%Ru alloy after fully heat treated.

Table 2. Concentration distribution of elements in γ/γ′ phases of various alloys.

Alloy	Region	Al	Ta	Cr	Co	Mo	W	Ru	Total
	γ phase	2.72	0.45	16.4	11.78	8.32	2.32	0	41.99
Alloy 1	γ' phase	18.91	3.72	1.68	3.00	1.53	0.81	0	29.65
	Ratio	6.95/1	8.28/1	1/9.74	1/3.92	1/5.44	1/2.85	-	1/1.42
	γ phase	4.69	0.49	15.12	10.75	7.13	2.27	3.13	43.58
Alloy 3	γ' phase	18.20	3.52	1.98	4.03	2.21	1.73	1.41	33.08
	Ratio	3.88/1	7.18/1	1/7.64	1/2.67	1/3.22	1/1.31	1/2.21	1/1.32

The results show that Ru can improve the solubility of elements and decrease the distribution ratio in γ and γ′ phases. It is concluded that the main difference between the two alloys lies in the presence or absence of Ru element. It can be inferred that the addition of Ru element is the main reason for the "reverse distribution" effect. At the same time, adding Ru can increase the total atomic fraction of elements in γ phase from 41.99% to 53.58%, and in γ′ phase from 29.65% to 33.08%. This indicates that Ru can improve the alloying degree of γ′ phase of alloy, which is also the main reason that alloy containing Ru has better creep properties under high temperature.

4. Conclusions

(1) The creep life of Ru/4%W alloy, Ru/6%W alloy and 2%Ru/6%W alloy is 42h, 98h and 182h respectively under 1100 C/137 MPa.

(2) W is one of the main forming elements of TCP phase. The increase of W content can promote the precipitation of TCP phase and lead to a sharp decrease in creep properties of the alloy.

(3) During high temperature creep, W in γ′ phase can migrate to γ matrix of free-Ru alloy, meanwhile, W has a larger concentration gradient in γ/γ' phases transition region. The distribution of W in γ/γ' phase is relatively smaller, and the concentration gradient in γ/γ' phase transition region is also small in 2%Ru alloy.

(4) The addition of Ru can make the alloy produce "reverse distribution" effect, which can improve the alloying of γ′ phase, and make the alloy containing Ru have better high temperature creep properties.

Acknowledgments

The authors would like to thank the support of the program for Excellent Talents of Science and Technology in Yingkou Institute of Technology (RC201908) and Liaoning Provincial Natural Science Foundation Guidance Plan named Study on the Mechanism of Ru/Re in Nickel-based Single Crystal Superalloy for financial supports.

References

[1] Kabanov I V, Lomberg B S and Sidorina T N. Development of superalloy production at the electrostal metallurgical plant[J]. *Metallurgist*, 2017, **61(7-8)**:565-568.

[2] Yu Z H, Liu L, Zhang J. Effect of carbon addition on carbide morphology of single crystal Ni-based superalloy[J]. *Transactions of Nonferrous Metals Society of China*, 2014, **24(2)**: 339-345.

[3] Cui C Y, Gu Y F, Ping D H, et al. Effects of Ru additions on the microstructure and phase stability of Ni-base superalloy, UDIMET 720LI[J]. *Metallurgical and Materials Transactions A*, 2006**(37A)**: 355-360.

[4] Tan X P, Liu J L, Jin T, et al. Effect of ruthenium on high-temperature creep rupture life of a single crystal nicker-base superalloy[J]. *Materials Science and Engineering A*, 2011 **(528)**: 8381-8388.

[5] Neumeier S, Ang J, Hobbs R A, et al. Lattice misfit of high refractory ruthenium containing nickel-base superalloys[J]. *Advanced Materials Research*. 2011 **(278)**: 60-65.

[6] Liu E Z, Guan X R. Effect of Ru on corrosion resistance of high Cr content superalloy[J]. *Materials and Corrosion*, 2016, **67(12)**:1269-1273.

[7] Chen J Y, Feng Q, Sun Z Q. Topologically close-packed phase promotion in a Ru-containing single crystal superalloy[J]. *Scripta Materialia*, 2010**(63)**: 795-798.

[8] Feng Q, Carroll L J, Pollock T M. Solidification segregation in ruthenium-containing nickel-base superalloys[J]. *Matallurgical and Materials Transactions A*, 2006**(37A)**: 1949-1962.

[9] O'Hara K S, Walston W S, Ross E.W, et al. Nickel-base superalloy has improvsd combination of stress-rupture life microstructural stability[J]. U.S. *Patent* : **5482789**. 1996.

[10] Ofori A P, Humphreys C J, Tin S, et al. A TEM study of the effect of platinum group metals in advanced single crystal nickel-based superalloys[J]. *Superalloys 2004*. Warrendale, PA: TMS, 2004, 787-794.

Micro-electrochemical evaluation of α- or β-type titanium alloys

Junhui Lin[1], Jiafei Lu[1], Zhenhua Dan[1 *], Yi Ding[1] and Hui Chang[1]

[1] College of Materials Science and Engineering and Tech Institute for Advanced Materials, Nanjing Tech University, Nanjing 210009, China

Abstract. In order to evaluate electrochemical stabilities of α, β phases, the micro-electrode techniques was employed to test the micropolarization behavior of α-type pure Ti and β-type Ti15Mo alloys. The comparison between macropolarization and micropolarization curves demonstrates that the α-type pure Ti alloy tends to exhibit the uniform corrosion, on the other hand, the pitting corrosion is the main form for the β-type Ti15Mo alloys. The α grains with different orientations shows similar polarization behavior, but the polarization behavior of those β grains are strongly affected by their crystalline orientations, particularly the worse electrochemical stabilities of β grains with (200)-dominance. The corrosion potential and corrosion current density of α-type pure Ti is lower than β-type Ti15Mo alloys. The α grains have better electrochemical stabilities in comparison to β grains, which indicate that the α+β dual-phase Ti-6Al-4V ELI alloys is highly possible to suffer from the preferential dissolution of β grains in free immersion condition and/or polarized states. Finally, by combining the SEM, EDS and polarization curves of dual-phase Ti-6Al-4V ELI alloys, it is concluded that β phases are corroded prior to α phases.

1. Introduction

Titanium alloys are widely used in the aerospace, marine engineering, biomedical and other fields because of their good properties such as good fatigue and corrosion resistance, high strength to weight ratio, non-magnetic and good biocompatibility[1-5]. Titanium alloys are generally classified into three types according to their different microstructures, α-type, β-type, and α+β titanium alloys[6]. Currently, the most widely used are α-type and α+β titanium alloys. With the extensive applications of titanium alloys in the ship industries and marine engineering, it is especially necessary to investigate the corrosion behavior of titanium alloys. Characteristic differences of α and β phases in dual-phase titanium alloys can result in the different corrosion behavior of α- and/or β-phase in the corrosive mediums, such as corrosion forms, corrosion morphology and even the corrosion mechanisms. In special corrosion conditions, α or β phases are preferentially corroded and become vulnerable regions of corrosion, resulting in the failures of titanium alloys in the form of stress cracking corrosion. Therefore, evaluation of the electrochemical properties of the α and β phases is important for understanding the corrosion mechanisms of titanium alloys. The macropolarization behavior of dual-phase titanium alloys has been widely evaluated by the conventional polarization measurements to reflect the integrated effects of all coexisting phases in the matrixs before. The capillary microelectrode methods have been developed to investigate the corrosion behavior of inclusions[7-9] and the intermetallic phases last decades[10,11]. However, the evaluation of the individual α phases and β phases in the electrochemical stabilities has not been conducted before.

In order to study the different electrochemical properties of α and β phases, α-type pure Ti, β-type Ti15Mo alloys and dual-phase Ti-6Al-4V ELI alloys were studied by traditional macropolarization electrochemical methods and micro-area masking corrosion testing techniques. The X-ray diffraction (XRD), energy dispersive X-ray spectroscopy (EDS), scanning electron microscopy (SEM), optical microscopy (OM) and micro-electrochemical experiments were employed to judge which has a better electrochemical stability between α and β phases of the titanium alloys. Through the above studies, micro-electrochemical evaluation of α, β phases of titanium alloys can be achieved.

2. Experimental procedure

Two model alloys, α-type pure Ti and β-type Ti15Mo alloys, were prepared by using a water-cooled cold copper crucible suspension melting furnace, and the samples of Ti-6Al-4V ELI alloy with basket-weave microstructure were purchased from Baotai Co. Ltd. Both α-type pure Ti and β-type Ti15Mo alloys were used in their as-forged states, because the phase size of the alloys was several hundredth micrometers so that the micropolarization behavior of two alloys can be easily studied by using the micro-area masking corrosion testing techniques. The specimens were polished successively to 2000 grits with SiC papers and 1 μm Al_2O_3 pastes. The as-forged alloys were directly used for electrochemical evaluation because the α or β grains are able to expose to the testing solutions by using the organic sealing chemical agents under the help of optical microscopes. Before the electrochemical and surface analysis tests, the samples were ultrasonically cleaned in ethanol and pure water. The etching solution of $HF:HNO_3:H_2O$ (volume ratio is set as 1:2:7) is used to prepare the metallographic samples. Electrochemical testing of two alloys were performed in the self-made electrochemical cells equipped with a reference electrode of Φ0.1 mm Ag wires with AgCl coatings and a counter electrode of Φ0.1 mm Pt wires. The test areas were masked by using the insoluble organic agents purchased from Wako Sendai Co. Ltd. Micro-area masking corrosion testing techniques[9] were used to perform the micro-polarization measurements on the exposed areas marked with aid of XRD analysis and optical microscope. The polarization curves were collected in 0.1mol/L NaCl solution by an electrochemical workstation (Corrtest, CS350) with a scan rate of 1 mV/s. The metallographic structure and microstructure of two alloys were observed by using an optical microscope (OM, ZEISS, AxioObserver.A1m) and a scanning electron microscope (JEOL, FEI Quanta 200). The chemical compositions of the samples were determined using an energy dispersive X-ray spectroscope (EDS, JEOL, FEI Quanta 200). Identification of the grain orientations and the crystalline states of the model alloys at the corresponding exposed areas were conducted by an X-ray diffractometer (XRD, Rigaku, Geigerflex) and an optical microscope (OM, ZEISS, AxioObserver.A1m).

3. Results

3.1. Micro-polarization behavior of α phases in α-type pure Ti

The metallographic photos of as-forged α-type pure Ti are shown in Fig 1. It can be seen that the metallographic structure of α-Ti consists of α phases with a size of several hundredth micrometers. The masked micro-regions with rectangular areas of about 100 μm×100 μm are marked in Fig 1. The α-type pure Ti has close-packed hexagonal structure. The XRD patterns of α-Ti macro-region and six micro-regions are shown in Fig 2. It can be clearly seen that the XRD patterns in macro-regions are typical and those in the micro-regions are different in the crystalline orientations. The diffraction peaks at 2θ of 35°, 38°, 40°, 53°, 63°, 70°, 76° and 77° were assigned to (100), (002), (101), (102), (110), (103), (112) and (201), which are corresponded to the diffraction patterns of α phases. Figure 3 shows the polarization curves of the macro-region and the corresponding micro-regions marked in Fig 1. The curve fluctuates around -0.1, 0, 0.05 and 0.1 V in the polarization curve of the macro-region which is considered to be due to the high dissolution of less stable α phases with different orientations such as (101) dominant grains in Region 2 and 4. The grains with a preferential dominance of (002) in Region 1, 3, 5 and 6 have relatively large passive current densities, indicating their higher electrochemical activities. The vulnerable parts of corrosion protective films on α-type pure Ti surface are damaged instantaneously while being pitted, and the corrosion protective films were repaired rapidly. It can be seen from the corresponding polarization curves that both corrosion potential and corrosion current density are larger than those in macro-region. It can be concluded that the corrosion potential and pitting potential of α-Ti are about -0.42 and 0.11 V, respectively. The corrosion current density of macro-region is about 2.65×10^{-6}A/cm^{-2} and the other six micro-regions are about 8.07×10^{-6} A/cm^{-2}. The electrochemical properties of the macro-region and the other six micro-regions are not much different.

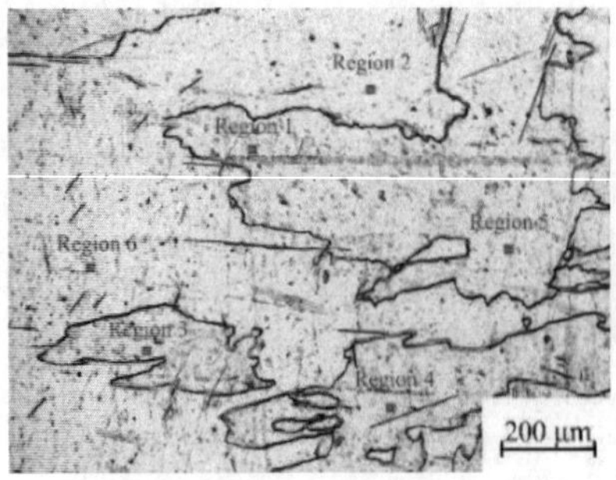

Fig.1 OM micrograph of α-type pure Ti in macro-region.

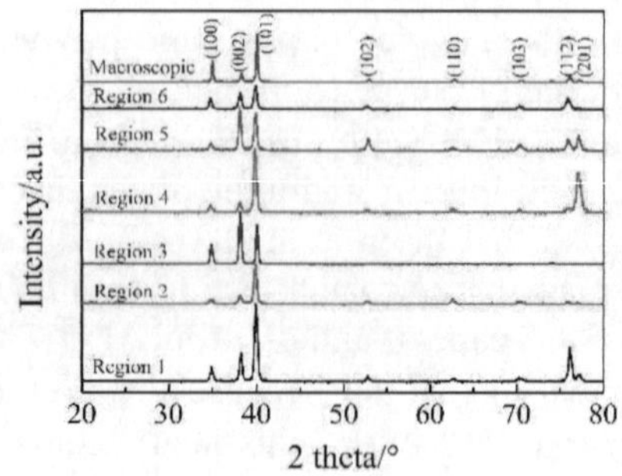

Fig. 2 XRD patterns of α-type pure Ti in macroscopic and different regions marked in Fig.1.

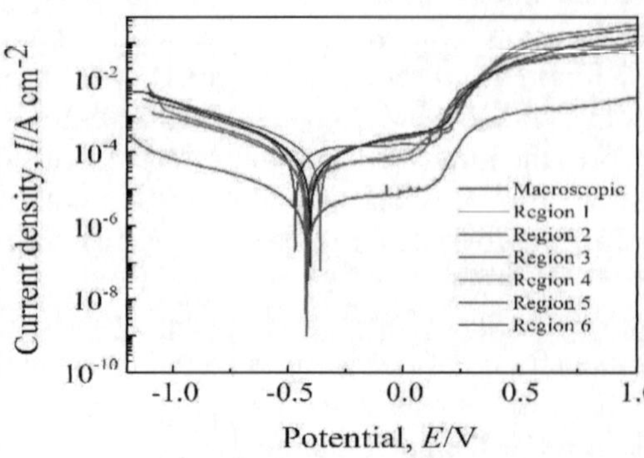

Fig.3 Polarization curves of α-type pure Ti in macroscopic and different micro-regions.

Fig. 4 shows the SEM morphology of α-type pure Ti in macroscopic and different grains before and after micro-electrochemical tests. There are almost no pits on the surface of the remaining micro-region in Fig. 4a. SEM images of the α-type pure Ti without masking shows the uniform corrosion without localized corrosion in the whole macro-region. The dissolution tended to be uniform. In other regions, such as the corrosion morphology in Region 3-5 in Fig. 4, the corrosion were considered to be uniform. But the corrosion morphology in Fig. 4b, several small pits can be observed which are considered to be from the some small intermetallic particles distributed from the interstitial inclusions. Therefore, the corrosion for α-type pure Ti is concluded to be uniform and the effect of the grain boundaries is able to be neglected.

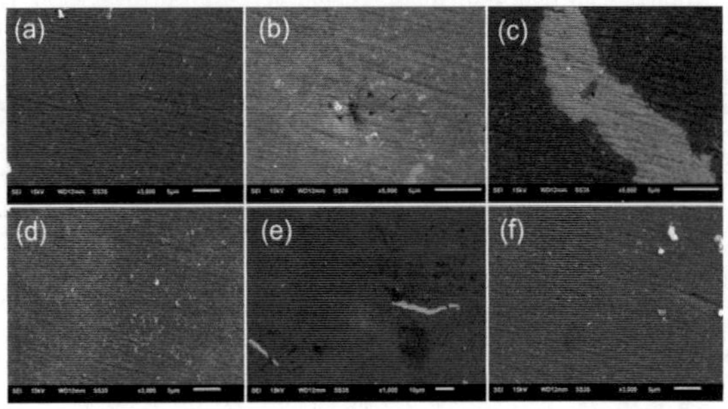

Fig.4 SEM morphology of α-type pure Ti in macroscopic (a) and marked micro-regions in Region 2 (b), Region 3 (c), Region 4 (c), Region 5 (d) and Region 6 (f).

3.2. Micro-polarization behavior of β-type Ti15Mo alloys

The metallographic structure of as-forged β-type Ti15Mo alloy is shown in Fig 5. The grains of β-type Ti15Mo alloys are several hundredth micrometers in size. The selected micro-regions on different grains are marked in Fig. 5. The β-type Ti15Mo has body-centered cubic structure. XRD patterns on the macro-region and six micro-regions are shown in Fig 6. Compared with complex orientations in the XRD patterns of α-type pure Ti, there are three orientations in the crystal planes of β-Ti(Mo) grains, which are (110), (200), (211) corresponding to the diffraction peaks at 2θ of 39°, 56° and 71°, respectively. In Region 1, the XRD pattern in Fig. 6 shows the dominant distribution of (200) and (110) orientations of the crystal planes. The (211) and (110) orientations of the crystal planes are prevailed in Region 5. The XRD patterns in other micro-regions are similar with that of macro-regions. The macropolarization curves of macro-region and the micropolarization curves of six micro-regions of the β-type Ti15Mo alloys are shown in Fig. 7. It can be seen that the corrosion potential in the macro-region and the other six micro-regions is between about -0.26 V. The corrosion potential of the macro-region is about -0.23 V, and its corrosion current density is smaller than those at the other six micro-regions. The breakdown potential of the native passive films, so-called pitting potential, was 0.13 V. Compared with the polarization curves of α-type pure Ti, the passivation potential range of β-type Ti15Mo alloy is narrow. The micropolariztion curves in the β grains can be divided into two groups: Group I from Region 1-3 and Group II from Region 4-6. It can be seen from the XRD pattern of Region 1 that the peak intensity of the (200) crystal plane is significantly higher than that of the macroscopic region. The pitting potential is confirmed to be -0.15 V, comparable to those in Region 2 and 3. The corresponding micro-polarization curve continuously fluctuates between -0.15 and 0.13 V. This is due to the repeated process of pitting corrosion destroying the protective film on the surface of the regions and the rapid self-healing of the protective films, which results in the increase of pit depths and corrosion area. As shown in Fig. 8b, a plurality of corrosion pits are joined together to form a large one. In addition, the micropolarization curves of Regions 2 and 3 are very similar to that of Region 1, and the corrosion behavior is considered to be similar. The corrosion potential is about -0.29 V, and the corrosion current density is significantly larger than that of the macro-region. The polarization curves of Group II for Region 4, 5 and 6 show that the corrosion potential is about -0.24 V, close to that of macro-region, and the corrosion current density is also larger than macro-region. Thus, the corrosion potential of β-Ti(Mo) phases is slightly positive than that of α-Ti phases, which is ascribed to integrated enhancement of the corrosion resistance of passive films consisting of MoO_x[12-14]. The corrosion current density of macro-region is about 6.59×10^{-6}A/cm^{-2} and the other six micro-regions are about 3.94×10^{-4} A/cm^{-2}. However, the corrosion current density of micro-regions is about two orders of magnitude larger than that of the α-Ti phases. On the basis of the micropolarization curves of Group I and Group II, the distribution of (110) and (200) orientations played an important role of the shift of the pitting potential. The more the grains with higher distribution of (110) and (200) orientations, the worse corrosion resistance is regarded to be.

Fig.5 OM micrograph of β-type Ti15Mo alloys in macro-region.

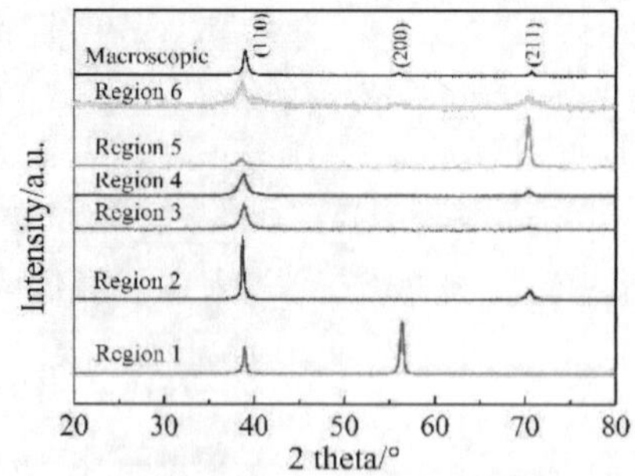

Fig.6 XRD patterns of β-type Ti15Mo alloys in macroscopic and different regions.

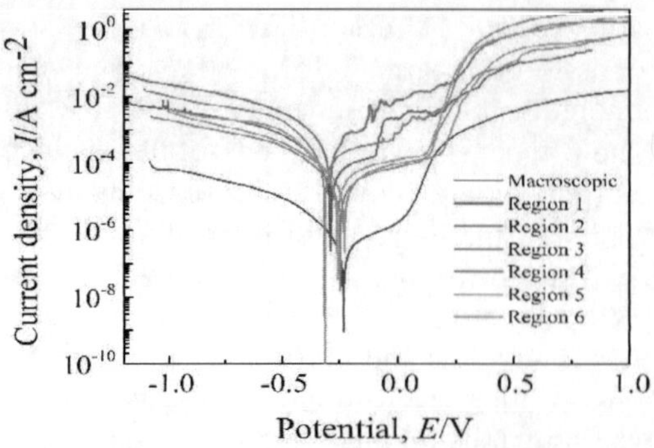

Fig.7 Polarization curves of β-type Ti15Mo alloys in macroscopic and different regions.

Figure 8a is corrosion morphology of as-forged β-type Ti15Mo alloy, indicating that the corrosion on as-forged β-type Ti15Mo alloy is more localized and many corrosion pits on the surface of the sample formed. SEM images in Fig. 8(b)-(f) also show that many pits with different sizes may be generated during anodic polarization. The corrosion morphology of Group I included many small sized corrosion pits which might be due to the lower electrochemical stabilities of the grains with (110) and (200) orientations of β-Ti(Mo) phases. The large corrosion pits formed on Group II is considered to be due to the distribution of the intermetallic particles in the matrix. The corrosion morphology in Fig. 8f could be a supportive evidence due to the existence of the single intermetallics particle inside the corrosion pits.

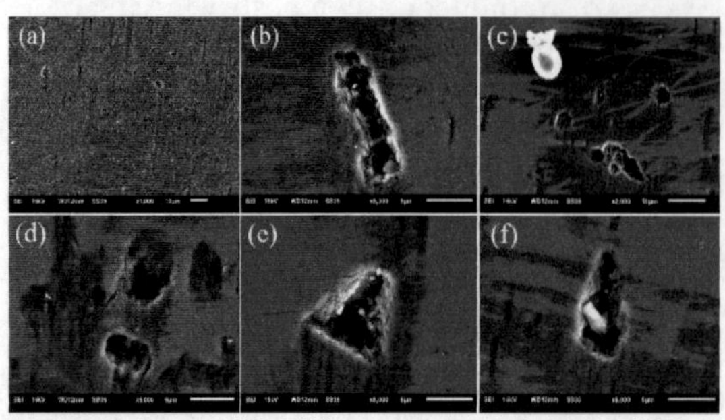

Fig.8 SEM morphology of β-Ti in macroscopic and different regions.

3.3. Integrated macro-polarization behavior of dual-phase Ti-6Al-4V ELI alloys

Figure 9a is a SEM image of α+β dual-phase Ti-6Al-4V ELI alloys after etching in the above metallographic solution. The black area in the image corresponds to α phase, and β phases in the flaky white area are distributed in Ti-6Al-4V ELI alloys with the basket-weave microstructure. After corrosion, Ti-6Al-4V ELI alloys structure has been difficult to see with basket-weave structure, because phase structure has been corroded in large quantities. However, the distribution tendency of the stabilizing Al element for α phase and V element for β phase is considered to be the similar. EDS spectra of Zone A and Zone B in Fig. 9b exhibited that the elemental composition of Zone A is 94.38% of Ti,4.65% of Al and 0.97% of V, and these of Zone B is 88.27% of Ti, 3.20% of Al, and 8.53% of V. The largest difference in the elemental composition between Zone A and B is the V content, and the V content in Zone A is significantly less than the percentage of Zone B. It can be seen that the major phases are α-Ti, and the white β phases have been dissolved preferentially. As shown in the polarization curve in Fig. 9c, there is several small increase of the current densities at 0.07-0.12V, which might be due to the preferential dissolution of the β phases in Ti-6Al-4V ELI alloys. On the other hand, there are many etched grain boundaries and the black regions in the metallographic images in Fig. 5, which is regarded to result from the fast corrosion of the β phases. These facts prove that the β phases in Ti-6Al-4V ELI alloys have higher electrochemical activities and is corroded prior to coexisting α phases[15]. The macropolarization behavior of single-crystal titanium alloys might be better for understanding the corrosion mechanism, although the experiments are difficult to conduct so far.

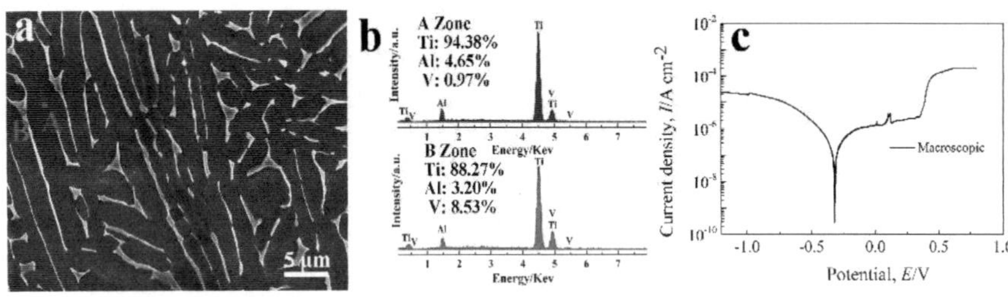

Fig. 9 Scanning electron microscope (SEM) images of α+β dual-phase Ti-6Al-4V ELI alloys (a), EDS spectra corresponding to Zone A and B (b), and Macro-polarization curves of Ti-6Al-4V ELI alloys in macro-region (c).

4. Summary

On basis of the crystalline and orientation analysis by XRD, observation of the corrosion morphology and the micropolarization behavior of the two α-type pure Ti and β-type Ti15Mo alloys and two-phase Ti-6Al-4V ELI alloys, the orientations of the crystalline planes and the distribution of the intermetallic phases can affect the micropolarization behavior. The corrosion of α-Ti phase in α-type pure Ti tends to take place uniformly, but that of β-phase in β-type Ti15Mo alloys is more localized. The corrosion rate of α phases in the 0.1 mol/L NaCl solution is significantly lower than that of β phases. For the two-phase titanium alloys, β phases are corroded prior to α phases due to the less electrochemical stabilities of β-phases.

Acknowledgement

The authors would like to acknowledge Jiangsu Collaborative Innovation Center for Advanced Inorganic Function Composites, State Key Lab of Advanced Metals and Materials under Grants No. 2018-ZD04, State Key Laboratory of Metal Material for Marine Equipment and Application (Grant No. HG-SKL(2018)06) and Priority Academic Program Development of Jiangsu Higher Education Institution (PAPD). This work is financially supported by the Natural Science Foundation of China under Grant No. 51671106, Natural Science Foundation of Jiangsu Province under Grants No. BK20171424 and National Defense Basic Scientific Research Program of China (Grant No. JCKY08414C020).

References

[1] Boyer R R and Briggs R D 2005 *J. Mater. Eng. Perform.* **14** 681-5
[2] Oryshchenko A S, Gorynin I V, Leonov V P, Kudryavtsev A S, Mikhailov V I and Chudakov E V 2015 *Appl. Res.* **6** 571-9
[3] Long M and Rack H J 1998 *Biomaterials.* **19** 1621-39
[4] Rack H J and Qazi J I 2006 *Mater. Sci. Eng. C.* **26** 1269-77

[5] Anasane S S and Bhattacharyya B 2017 *J. Manuf. Process.* **28** 285-294
[6] Veiga C, Davim J P and Loureiro A J R 2012 *Rev. Adv. Mater. Sci.* **32** 133-148
[7] Suter T and Böhni H 1997 *Electrochim. Acta.* **42** 3275-80
[8] Suter T, Peter T and Böhni H 1995 *Mater. Sci. Forum* **192** 25-40
[9] Muto I, Kurokawa S and Hara N 2009 *ECS Trans.* **16** 269-279
[10] Birbilis N and Buchheit R G 2005 *J. Electrochem. Soc.* **152** B140-151
[11] Birbilis N and Buchheit R G 2008 *J. Electrochem. Soc.* **155** C117-126
[12] Nishimura T 2011 *Nucl. Eng. Des.* **241** 4745-49
[13] Park P Y, Akiyama E, Habazaki H, Kawashima A, Asami K and Hashimoto K 1996 *Corros. Sci.* **38** 1649-67
[14] Nogueira R A, Pinto L M C, Ângelo A C D, Claro A P R A and Grandini C R 2013 *Mater. Res.* **16** 1398-1404
[15] Atapour M, Pilchak A L, Frankel G S and Williams J C 2011 *Mater. Sci. Eng. C.* **31** 885-891

Institute of Physics
Dirac House, Temple Back
Bristol BS1 6BE UK

ISSN: 1757-8981
ISBN 978-1-5108-9840-0

CURRAN ASSOCIATES INC.
proceedings.com